SCIENTIFIC FACTS IN THE BIBLE

Impeccable Evidence the Bible is Infallible

Dr. Maxwell Shimba

Copyright © 2024 – Dr. Maxwell Shimba

All rights reserved. No portion of this book may be reproduced, stored in a retrieval system, or transmitted in any form or by any means – electronics, mechanical, photocopy, recording, scanning, or other – except for brief quotations in critical reviews or articles, without the prior written permission of the publisher.

Shimba Publishing, LLC.

Printed by Shimba Publishing LLC
Printed in the United States of America

TABLE OF CONTENTS

DID GOD CREATE MATHEMATICS?

The concise answer to whether God created Mathematics is an affirmative, yes. Colossians 1:16 tells us, "For by Him all things were created, in heaven and on earth, visible and invisible... all things were created through Him and for Him." The term 'all' encompasses Mathematics, among the myriad of other creations by God.

Mathematics traces its roots back to the book of Genesis, where God numbered the days of Creation. Man did not invent Mathematics but rather developed symbols, numbers, and equations to apply and understand God's creation of Mathematics. Only a perfect, all-knowing God could craft something as intricate, logical, and consistent as math, which inherently demands order, a concept created by God. As one mathematician aptly stated, "He allows us to explain His order with equations that always have the same answer."

The relationship between God and math is profound. If we acknowledge that God created Mathematics, we can contemplate why He did so and how it aids in our relationship with Him. Primarily, God created math, like the universe and all within it, including us, for His glory. Math illuminates God's sovereignty, infinity, and perfect wisdom. When we delve into the realm of Mathematics, we cannot help but marvel at its Creator, who fashioned it to be unfailingly consistent and reliable. We employ math to solve problems, construct, surmount obstacles, and manage our finances in a manner that glorifies God.

Moreover, math facilitates our relationship with God by integrating logic and order into our daily lives. Math equips us with solutions to everyday challenges and problems. We utilize it to reconcile accounts, settle debts, follow recipes, conduct business, and undertake household repairs and enhancements.

Mathematics is also a testament to God's faithfulness. Our reliance on math is grounded in the fact that our faithful, all-powerful God consistently upholds our universe (Colossians 1:17). He ensures that the principles of addition, subtraction, multiplication, and division unfailingly yield clear outcomes. We can trust that since God is flawless and impeccably created all things, He ensures that math operates flawlessly.

The Bible contains over 150 references to math, demonstrating that the principles of addition, subtraction, multiplication, and division are not merely mathematical but also spiritual.

Addition

The principle of addition extends beyond adding numbers; believers are instructed in 2 Peter 1:5–7 to add virtues to their faith. They are to add virtue to knowledge, self-control to perseverance, and godliness to brotherly kindness, ultimately culminating in love.

Subtraction

A remarkable example of subtraction in scripture is found in Judges 7, where Gideon's army is reduced from 32,000 to just 300 men by God's instruction. Despite the drastic reduction in numbers, God granted them victory over the Midianites, showcasing His power over human limitations.

Multiplication

One of Jesus' notable miracles involved multiplication, as recounted in Matthew 14:13–21, where he fed thousands with just five loaves and two fish. This miracle not only satisfied the hunger of the multitude but also demonstrated Jesus' divine power.

Division

In Ezekiel 47, God instructed the prophet to divide the land of Israel among the twelve tribes. Additionally, the division of Jesus' garments among the soldiers at His crucifixion fulfills a prophecy and underscores the significance of division in biblical contexts.

In conclusion, Mathematics is not a human invention but a divine creation designed by God for His glory and our benefit. It serves as a testament to God's wisdom, consistency, and faithfulness, highlighting His role as the ultimate mathematician and the foundation of all logic and order.

DR. MAXWELL SHIMBA

CHAPTER 01

JESUS MATHEMATICAL MIRACLE: FEEDING THE MULTITUDE WITH FACTORIAL NOTATION

Factorial notation, denoted by the symbol "!", is a mathematical notation used to represent the product of all positive integers up to a given non-negative integer. For example, the factorial of 5 is written as "5!" and is calculated as:

5! = 5 times 4 times 3 times 2 times 1 = 120.

In general, the factorial of a non-negative integer *n*, denoted as *n!*, is the product of all positive integers less than or equal to *n*. This can be expressed as:

n! = n times (n-1) times (n-2) times … times 2 times 1.

Factorials are commonly used in combinatorial mathematics to calculate permutations and combinations, as well as in various other mathematical and scientific contexts.

Jesus' Mathematical Miracle: Feeding the Multitude with Factorial Notation

The feeding of the 5000 is one of the most famous miracles of Jesus, demonstrating His compassion and divine power. While the Bible does not explicitly mention factorial notation in this event, we can explore how this mathematical concept aligns with the miracle of multiplying the loaves and fishes.

Understanding Factorial Notation

Factorial notation, represented by the symbol "!", is used to calculate the product of all positive integers up to a given number. For example, **5! = 5 times 4 times 3 times 2 times 1 = 120.**

The Miracle of Multiplication

In the Gospel accounts of Matthew, Mark, Luke, and John, Jesus fed a multitude of people with just five loaves of bread and two fish. Let's consider how factorial notation could represent this miracle:

The story of Jesus feeding the 5000 with just 7 pieces of food (5 loaves of bread and 2 fish) is a miracle recorded in the Bible in the Gospels of Matthew, Mark, Luke, and John. Mathematically, this miracle is often viewed in terms of multiplication, not factorial notation. However, we can explore how factorial notation might conceptually relate to this miracle.

Factorial notation is typically used to calculate the number of ways to arrange a certain number of items. In the case of the feeding of the 5000, we can think of factorial notation in a more abstract way, considering the miracle as a demonstration of God's power to multiply the food.

Let's say each piece of bread and fish is represented by a unique item (e.g., Bread1, Bread2, Fish1, Fish2). The factorial of 7 (7!) would represent all the possible arrangements of these items:

7! = 7 times 6 times 5 times 4 times 3 times 2 times 1 = 5040.

In this scenario, 5040 would represent the theoretical maximum number of ways these 7 pieces of food could be arranged. However, in the miracle, Jesus did not arrange the food in different ways but rather multiplied it to feed the 5000, illustrating that with God, all things are possible, including multiplying food to meet the needs of many. This is a demonstration of God's power and provision, which goes beyond simple mathematical calculations.

Significance of the Miracle

This mathematical perspective highlights the miraculous nature of Jesus' provision. In a natural sense, it would be impossible to feed thousands of people with such a small amount of food. However, through His divine power, Jesus not only provided enough food but also had an abundance left over.

While factorial notation is a mathematical concept, its application to the feeding of the 5000 serves to deepen our understanding of the magnitude of Jesus' miracle. It underscores His ability to transcend natural laws and provide for our needs abundantly. As we reflect on this miracle, may it strengthen our faith in the One who can multiply our meager offerings into abundance.

Symbolism and Spiritual Meaning

The feeding of the 5000, while a miraculous event, also holds profound symbolism and spiritual meaning. Beyond the mathematical implications, this miracle teaches us valuable lessons about faith, provision, and the nature of Jesus.

Symbolism of the Bread

Bread holds significant symbolism in the Bible, often representing sustenance, provision, and the Word of God. In this miracle, the five loaves of bread may symbolize the five books of the Torah, the foundational teachings of the Old Testament. Just as Jesus provided physical nourishment through the bread, He also provides spiritual nourishment through His teachings.

Symbolism of the Fish

Fish, in biblical symbolism, can represent abundance, multiplication, and discipleship. The two fish in this miracle may symbolize the Old and New Testaments, working together to bring about the multiplication of spiritual blessings.

Multiplication and Abundance

The act of multiplying the loaves and fishes to feed the multitude demonstrates Jesus' ability to provide abundantly. It reminds us that God is not limited by our resources or circumstances. Just as He multiplied the food to feed the crowd, He can multiply our efforts, resources, and blessings to meet our needs.

Faith and Provision

The miracle also highlights the importance of faith and trust in God's provision. When faced with the impossibility of feeding so many with so little, Jesus demonstrated that nothing is impossible with God. The disciples' faith was tested, but through this miracle, they witnessed firsthand the power and provision of Jesus.

Lessons for Today

The feeding of the 5000 continues to offer lessons for us today. It reminds us to trust in God's provision, even when faced with scarcity. It challenges us to be generous with what we have, knowing that God can multiply our offerings for His glory. It also encourages us to rely on Jesus as the bread of life, who satisfies our spiritual hunger and sustains us through all circumstances.

As we reflect on the feeding of the 5000, may we be reminded of God's faithfulness, provision, and abundance in our lives. May we trust Him to multiply our efforts and resources for His purposes, knowing that He can do far more abundantly than all we ask or imagine.

Sure, here's the next chapter continuing the exploration of how Jesus used mathematical concepts to perform the miracle of feeding 5000 people with 7 pieces of food:

Divine Multiplication and Miraculous Provision

The Miracle of Multiplication

The miracle of Jesus feeding the 5000 with five loaves of bread and two fish, recorded in all four Gospels (Matthew 14:13-21, Mark 6:30-44, Luke 9:10-17, John 6:1-14), is a profound example of divine multiplication. Unlike factorial notation, which deals with the arrangement and combination of items, the multiplication performed by Jesus transcends natural laws and human understanding.

Concept of Divine Multiplication

In mathematical terms, multiplication is a fundamental operation where a number is added to itself a certain number of times. For example, (5 times 2) means adding 5 twice, resulting in 10. However, Jesus' multiplication was not limited by natural constraints. He took the available resources and supernaturally increased their quantity to meet the needs of the people present. This act can be seen as an exponential increase rather than a simple multiplication.

Symbolism and Lessons

1. Resourcefulness and Faith: Jesus' miracle teaches about the importance of resourcefulness and faith. Even with limited resources (7 pieces of food), Jesus demonstrated that faith in God can lead to miraculous outcomes. The key takeaway is that God can use whatever little we have to achieve great things.

2. Provision and Abundance: The miracle also symbolizes God's provision and abundance. Not only were 5000 men, along with women and children, fed but there were also 12 baskets of leftovers, illustrating that God's blessings are more than sufficient.

3. Divine Order and Organization: Before performing the miracle, Jesus instructed the crowd to sit down in groups (Mark 6:39-40). This act of organizing the people highlights the importance of order and preparation in receiving God's blessings.

The Mathematical Perspective

While the miracle cannot be fully explained by human mathematics, we can use mathematical concepts to appreciate the scale of the miracle. If we consider the initial 7 pieces of food as a starting point, and assume each piece needed to be multiplied to feed each individual present, we see an extraordinary scale of multiplication.

If each of the 7 pieces of food was to be divided among the 5000 men:

$$\{Number\ of\ pieces\ per\ person\} = \frac{7}{5000}\ approximately \approx 0.0014$$

However, through Jesus' divine intervention, each of the 5000 people not only received enough to eat but also had leftovers, showing an exponential rather than linear multiplication.

Spiritual Multiplication

In a spiritual sense, the multiplication performed by Jesus can be seen as an invitation to trust in God's limitless power and provision. Just as the 7 pieces of food were multiplied to feed thousands, our faith and small acts of obedience can be multiplied by God to achieve extraordinary outcomes.

Reflection

The miracle of feeding the 5000 invites us to reflect on the nature of God's provision and our role in it. While human understanding and mathematics provide a framework for understanding the world, miracles remind us of the divine power that surpasses all human knowledge. By trusting in God and offering what little we have, we open ourselves to the miraculous possibilities of divine multiplication.

In conclusion, Jesus' use of multiplication in feeding the 5000 is a testament to God's ability to transcend natural laws and provide abundantly. This chapter serves as a reminder that faith in God's provision can lead to miraculous outcomes, far beyond what human mathematics can explain.

CHAPTER 02

HOW COMPLEX IS TRINITY? 1 + 1 + 1= 3

What is Ternary Addition?

Ternary addition is a form of arithmetic used in a base-3 numeral system. Unlike the more familiar base-10 system, which uses digits from 0 to 9, the ternary system uses only three digits: 0, 1, and 2. Here's a basic introduction to ternary addition:

Ternary Digits

· 0

· 1

· 2

Addition Rules

When adding two ternary digits, the result might require a carry, just like in base-10 addition. Here are the rules for adding ternary digits:

1. $(0 + 0 = 0)$

2. $(0 + 1 = 1)$

3. $(0 + 2 = 2)$

4. $(1 + 0 = 1)$

5. $(1 + 1 = 2)$

6. $(1 + 2 = 10_3$ (which is equivalent to $1 \cdot 3^1 + 0 \cdot 3^0 = 3_{10}$); thus, write down 0 and carry 1)

7. $(2 + 0 = 2)$

8. $(2 + 1 = 10_3$ (same as above, write down 0 and carry 1)

9. $(2 + 2 = 11_3$ (which is equivalent to $1 \cdot 3^1 + 1 \cdot 3^0 = 4_{10}$; thus, write down 1 and carry 1)

Examples of Ternary Addition

1. Simple Addition Without Carry:

· $(1 + 1 = 2)$

· $(2 + 0 = 2)$

2. Addition With Carry:

· $(1 + 2 = 10_3)$ (write down 0, carry 1)

· $(2 + 2 = 11_3)$ (write down 1, carry 1)

Multi-Digit Ternary Addition

Let's add two ternary numbers, 12_3 and 21_3:

1. Start with the rightmost digit:

· $2 + 1 = 10_3$ (write down 0, carry 1)

2. Move to the next digit to the left, including the carry:

· $(1 + 2 + 1$ (carry) $= 11_3)$ (write down 1, carry 1)

So, the result is:

12_3

$+ 21_3$

110_3

Thus, $12_3 + 21_3 = 110_3$

Converting Between Bases

To fully understand ternary addition, it helps to convert between base-10 and base-3. Here's a quick guide:

1. To convert from base-10 to base-3:

- Divide the number by 3, record the remainder.

- Divide the quotient by 3, and record the remainder.

- Repeat until the quotient is 0.

- The base-3 number is the remainder read from bottom to top.

2. To convert from base-3 to base-10:

- Multiply each digit by (3) raised to the power of its position (starting from 0 on the right).

- Sum the results.

Example: Converting 110_3 to base-10:

$$1 \cdot 3^2 + 1 \cdot 3^1 + 0 \cdot 3^0 = 1 \cdot 9 + 1 \cdot 3 + 0 \cdot 1 = 9 + 3 +$$

$$0 = 12_{10}$$

Conversely, converting 12_{10} to base-3:

1. 12 divide 3 = 4 with a remainder of 0.

2. 4 divide 3 = 1 with a remainder of 1.

3. 1 divide 3 = 0 with a remainder of 1.

So, $12_{10} = 110_3$

Ternary addition, while different from the decimal system, follows a logical pattern and provides an interesting way to explore alternative numeral systems.

TRINITY: VERIFIED BY SCRIPTURES AND MATHEMATICS

Let's Start by Reading the Scripture:

1 John 5:7

For there are three that bear witness in heaven: the Father, the Word, and the Holy Spirit; and these three are one.

The Bible Verse States:

There are Three:

$$1 + 1 + 1 = 3$$

1. The Father

2. The Word - Jesus

3. The Holy Spirit

These three testify in heaven.

These Three are One:

$1 \cdot 1 \cdot 1 = 1$

God is Eternal: ∞

Psalm 90:2 tells us about the eternal nature of God:

Before the mountains were born or you brought forth the whole world, from everlasting to everlasting you are God.

Since humans measure everything by time, it is very difficult for us to conceive of something that has no beginning but has always been and will continue forever.

1. The Father is Eternal: He has no beginning or end ∞

2. The Son is Eternal: He has no beginning or end ∞

3. The Holy Spirit is Eternal: He has no beginning or end ∞

Jesus is Eternal ∞

Jesus Christ, God in the flesh, also confirmed His divinity and eternal nature to the people of His day by saying, "Before Abraham was born, I am" (John 8:58). Clearly, Jesus was claiming to be the eternal God in the flesh, as the Jews, upon hearing this, tried to stone Him. To the Jews, claiming to be the eternal God was blasphemy and worthy of death (Leviticus 24:16). Jesus was claiming to be eternal, just as His Father is eternal. The apostle John also declared this truth about Christ's nature: "In the beginning was the Word, and the Word was with God, and the Word was God" (John 1:1). Jesus and His Father are one, eternally existent, and share equally in the attribute of eternity.

The Holy Spirit is Eternal ∞

The Spirit of God/Lord/Christ: (Matthew 3:16, 2 Corinthians 3:17, 1 Peter 1:11) These names remind us that the Spirit of God is truly part of the Holy Trinity and that He is equal with God as the Father and the Son. He is revealed first at creation when He was "hovering over the waters," indicating His role in creation, along with Jesus who "made all things" (John 1:1-3). We see this Trinity of God again at Jesus' baptism, where the Spirit descended on Jesus and the voice of the Father was heard.

Now Let's Do the Mathematics:

Eternity in Mathematics or Arithmetic is represented by the symbol ∞, meaning "INFINITY."

Using ordinary mathematics, you can see:

$$\infty + \infty + \infty = \infty$$

Father ∞ + Son ∞ + Holy Spirit $\infty = \infty$

$1 + 1 + 1 = 3$ (There are three that bear witness in heaven)

$1 \cdot 1 \cdot 1 = 1$ (These three are one)

$$\infty + \infty + \infty = \infty$$

- God the Father has no beginning or end: The Father = ETERNAL = ∞

- The Word (Jesus) has no beginning or end: The Word "Jesus" = ETERNAL = ∞

- The Holy Spirit has no beginning or end: The Holy Spirit = ETERNAL = ∞

Reaffirming the Scripture:

1 John 5:7

For there are three that bear witness in heaven: the Father, the Word, and the Holy Spirit; and these three are one.

Applying Ternary Addition to the Trinity:

In the case of the Trinity, it involves the number three, and thus its rule must be of "TERNARY ADDITION" and not "BASE TEN ADDITION" as many use without realizing they are using the "BASE TEN" addition rule.

1. Ternary number uses BASE 3, unlike BASE 10 where when you count and reach 9, you revert to 1 and add "ZERO" = 10. In Ternary addition, when you count the final number is TWO, and the next number is 10, similar to "BINARY ADDITION" where your final number is ONE.

Proving the Trinity with Ternary Addition:

Using the ternary system:

A. $(1 + 1 + 1)_3 = 10_3$

You can see the answer is 10_3. Now, these numbers are reduced to ternary 10, which means the next step is to sum those digits "ONE and ZERO" using the same ternary addition base.

B. $(1 + 0)_3 = 1_3$

So, we have confirmed that the Trinity is verifiable using mathematics. Today, I have answered those who may have had doubts by proving that the Trinity is even answerable using mathematics.

Ask Your Math Instructor the Following:

A. Ternary of $(1 + 1 + 1) = ?)$

B. Ternary of $(1 + 2) = ?)$

After being answered, ask your instructor again:

C. Ternary of $(1 + 0) = ?)$

Indeed, God is the originator of mathematics, and there is nothing He cannot do.

God bless you all, and today you have been shown that the Trinity is verifiable even in mathematics.

Satisfying What Has Been Revealed:

A. The Father is called God (1 Corinthians 8:6).

B. The Son (Jesus) is called God (Isaiah 9:6; John 20:26-29).

C. The Holy Spirit is called God (Acts 5:3-4).

In Isaiah 48:16 and 61:1, the Son speaks while making reference to the Father and the Holy Spirit. Compare Isaiah 61:1 with Luke 4:14-19 to see the Son speaking. Matthew 3:16-17 describes Jesus' baptism. Here, God the Holy Spirit is descending on God the Son while God the Father declares His pleasure in the Son. Matthew 28:19 and 2 Corinthians 13:14 are examples of the three Persons in one God.

The Triune God and Mathematical Proof of Unity (1 John 5:7)

The concept of the Trinity, encapsulating the Father, the Son, and the Holy Spirit as one God, is a foundational doctrine in Christianity. While this mystery transcends human comprehension, mathematical principles can offer a way to visualize the unity and co-equality of the three Persons in the Godhead. This chapter explores how the Trinity can be understood mathematically, providing a unique perspective on the divine nature described in 1 John 5:7.

The Biblical Foundation

1 John 5:7 states:

"For there are three that bear witness in heaven: the Father, the Word, and the Holy Spirit; and these three are one."

This verse highlights the three distinct Persons who collectively constitute the one true God. The doctrine of the

Trinity asserts that these three Persons are co-equal, co-eternal, and consubstantial, meaning they share the same essence.

Mathematical Representation of the Trinity

Three Persons: Father, Son, and Holy Spirit

The Trinity can be mathematically represented using the concept of multiplication, where each Person of the Godhead is represented by the number one. This is because each Person is fully God, and multiplying them together reflects their unity.

$$1 \cdot 1 \cdot 1 = 1$$

Understanding the Multiplication

1. Father (1):

- The Father is God, fully and completely. Representing the Father by the number one signifies His complete divinity.

2. Son (1):

- The Son, Jesus Christ, is also fully God. Representing the Son by the number one signifies His complete divinity, equal to the Father.

3. Holy Spirit (1):

- The Holy Spirit is fully God as well. Representing the Holy Spirit by the number one signifies His complete divinity, equal to the Father and the Son.

When these three Persons are multiplied together, the result is still one, illustrating the profound unity within the Trinity:

$$1 \cdot 1 \cdot 1 = 1$$

The Unity and Co-Equality of the Trinity

Unity Through Multiplication

The mathematical operation of multiplication reflects the indivisible unity of the Trinity. Each Person is fully God, and their unity is perfectly represented by the multiplication of three ones resulting in one. This signifies that while the Father, Son, and Holy Spirit are distinct Persons, they are not separate gods but one God.

Co-Equality Through Identity

Using the number one for each Person underscores their co-equality. No Person in the Trinity is greater or lesser than another; they are all equal in essence and power. This is crucial for maintaining the balance of the Trinitarian doctrine.

Further Mathematical Considerations

Infinity in the Trinity

Another way to view the Trinity is through the concept of infinity, as God is infinite and eternal. Representing each Person as infinity ∞ emphasizes their eternal nature:

$$\infty \cdot \infty \cdot \infty = \infty$$

This further illustrates that the infinite nature of God is fully present in each Person of the Trinity, and their unity results in one infinite God.

Ternary Logic and the Trinity

Ternary logic, a system based on three values, can also be applied to understand the Trinity. In ternary addition, the sum of three ones $(1 + 1 + 1)$ in base-3 results in three, but in the divine context, this sum must be understood in light of their unified essence:

$$1 + 1 + 1 = 3$$

$$3 \rightarrow 1 \text{ (in the divine essence)}$$

This implies that while there are three distinct Persons, they operate as one single essence.

Theological Implications

Divine Mystery and Human Understanding

While mathematics provides a way to visualize the Trinity, it is ultimately a divine mystery that transcends human logic. The mathematical representation is a tool to aid understanding but does not fully encapsulate the depth of the divine nature.

Faith and Reason

The interplay between faith and reason is evident in the doctrine of the Trinity. Faith accepts the divine revelation of the Trinity, and reason seeks to understand it. Mathematics serves as a bridge, offering a logical framework to grasp this profound truth.

The mathematical proof of the Trinity, using multiplication to represent the unity of the Father, Son, and Holy Spirit, provides a unique perspective on this foundational Christian doctrine. By representing each Person as the number one, we can see how the multiplication of these three results in one, reflecting their unity and co-equality. This chapter underscores the importance of both faith and reason in understanding the divine mysteries and highlights the role of mathematics as a tool to aid our comprehension of the infinite and eternal nature of God.

CHAPTER 03

THE FEEDING OF THE 4,000

The miracle of the feeding of the 4,000, as recorded in Mark 8:1-9, is a profound testament to Jesus' compassion, provision, and divine power. This miracle, distinct from the feeding of the 5,000, further illustrates the theme of divine multiplication and the sufficiency of God's provision for His people.

The Setting

The event takes place in a remote region, where a large crowd has gathered to hear Jesus teach. The people have been with Him for three days, and their food supplies have run out. Jesus, seeing their need and not wanting to send them away hungry, initiates another miraculous feeding.

Initial Quantity: 7 Loaves and a Few Small Fish

The disciples present Jesus with what little they have: seven loaves of bread and a few small fish. This modest amount is insufficient by any human standard to feed such a large crowd. Yet, in the hands of Jesus, these meager resources are more than enough.

Jesus' Compassion and Command

Moved by compassion, Jesus instructs the crowd to sit down on the ground. He takes the seven loaves, gives thanks, breaks them, and gives them to His disciples to distribute. He does the same with the small fish. The act of giving thanks before the miracle highlights the importance of gratitude and trust in God's provision.

The Miracle: Feeding 4,000 Men

As the food is distributed, a miraculous multiplication occurs. The seven loaves and few fish are transformed into a feast sufficient for 4,000 men, not counting women and children. Everyone eats and is satisfied, a clear indication of the abundance and sufficiency of God's provision.

Result: 7 Baskets of Leftovers

After everyone has eaten their fill, the disciples gather the leftovers, filling seven baskets. The leftover food signifies not only the abundance of Jesus' provision but also the importance of stewardship and gratitude for God's gifts. It reinforces the message that with God, there is always more than enough.

Theological Significance

This miracle, like the feeding of the 5,000, reveals several key theological themes:

1. Jesus' Divine Authority and Compassion: Jesus' ability to multiply the loaves and fish demonstrates His divine authority over creation. His compassion for the physical needs of the crowd shows His deep love and care for humanity.

2. Faith and Obedience: The disciples' role in distributing the food illustrates the importance of faith and obedience. Despite their initial doubts, they trust Jesus' command and witness the miraculous provision.

3. God's Abundance: The seven baskets of leftovers underscore the theme of God's abundance. God provides not just enough, but more than enough, for those who trust in Him.

4. Spiritual Nourishment: Beyond the physical sustenance, this miracle points to the spiritual nourishment that Jesus provides. Just as He satisfies physical hunger, He also fulfills spiritual hunger, offering eternal life to all who come to Him.

Comparison with the Feeding of the 5,000

While similar to the feeding of the 5,000, this miracle has its unique elements. The different numbers involved (4,000 people, 7 loaves, 7 baskets) and the different locations highlight the recurring but distinct nature of Jesus' miracles. Both miracles together emphasize the widespread and inclusive nature of Jesus' ministry.

The feeding of the 4,000 is a powerful reminder of Jesus' ability to meet our needs abundantly. It calls us to trust in His provision, to act in faith, and to be grateful stewards of His gifts. This miracle reassures us that no matter how limited our resources may seem, in Jesus' hands, they can be transformed into an overflow of blessings.

Factorial notation is typically used in combinatorics and mathematics to represent the product of an integer and all the positive integers below it (e.g., $n! = n$ times $(n-1)$ times $(n-2)$ times 1. In the context of the feeding of the 4,000 with 7 loaves and a few small fish, factorial notation doesn't directly apply in a mathematical sense. However, we can draw a symbolic parallel to illustrate the concept of divine multiplication.

Symbolic Representation

Let's consider the story symbolically, using factorial notation as a metaphor for the miraculous multiplication that Jesus performed:

1. Initial Resources:

   - 7 loaves of bread

- A few small fish (let's assume 3 for this example, to have a number to work with)

2. Factorial Growth Concept:

- In a factorial sense, the multiplication of resources could be seen as a rapid and extensive growth, similar to how factorial numbers grow very quickly.

· 7! = 7 times 6 times 5 times 4 times 3 times 2 times 1 = 5,040

- While this is a large number, it symbolically represents the idea of abundance and rapid multiplication beyond normal expectations.

3. Miraculous Multiplication:

- Jesus took the 7 loaves and a few fish and multiplied them to feed 4,000 men, plus women and children. This multiplication was miraculous and far exceeded what could be explained by natural means.

4. Symbolic Parallel:

- Although not mathematically precise, using a factorial notation can help us understand the concept of exponential growth and divine multiplication.

- For example, if we consider the 7 loaves as having the potential for 7! (factorial) multiplication, it symbolically represents the idea that Jesus' miracle was an exponential increase in provision.

Illustrative Example

To illustrate this, let's create a symbolic mathematical metaphor:

- Starting with 7 loaves: Consider each loaf having the potential to be multiplied factorially.

- Factorial Growth:

- If we take 7 loaves and consider the potential for 7! we get 5,040. This isn't a literal number of loaves but represents the idea of abundant multiplication.

Mathematical Miracles

In a purely illustrative and symbolic sense, the idea of factorial growth can be used to represent the miraculous nature of Jesus' multiplication of food. Here's a step-by-step metaphorical breakdown:

1. Initial Quantity:

- 7 loaves of bread

- A few small fish (let's assume 3)

2. Divine Multiplication:

- Just as factorial numbers grow rapidly, Jesus' multiplication of the loaves and fish was a rapid, divine increase that fed thousands.

3. Result:

- Feeding 4,000 men plus women and children, with 7 baskets of leftovers, symbolizes the abundance and overflow resulting from divine intervention.

While factorial notation is not literally involved in the feeding of the 4,000, it can serve as a powerful metaphor for understanding the concept of divine multiplication. The miracle demonstrates Jesus' ability to take limited resources and multiply them far beyond natural expectations, providing an overflow of blessings and meeting the needs of thousands. This symbolic use of factorial notation helps us grasp the extraordinary nature of this biblical event.

CHAPTER 04

THE MIRACLE OF THE WIDOW'S OIL

The miracle of the widow's oil, found in 2 Kings 4:1-7, is a profound story of divine provision and faith. This miracle, performed by the prophet Elisha, illustrates the power of God's intervention in times of desperate need and demonstrates how faith and obedience can result in a continuous supply of God's blessings.

The Plight of the Widow

The story begins with a widow in dire straits. Her husband, a man who feared the Lord, had passed away, leaving her and her two sons in debt. The creditors were coming to take her sons as slaves to settle the debt. In her desperation, the widow turned to Elisha for help, crying out for assistance.

2 Kings 4:1 states:

"The wife of a man from the company of the prophets cried out to Elisha, 'Your servant my husband is dead, and you know that he revered the Lord. But now his creditor is coming to take my two boys as his slaves.'"

Elisha's Instructions

Elisha's response to the widow was practical yet rooted in faith. He asked her what she had in her house. The widow replied that she had nothing except a small jar of oil. This small jar of oil would become the centerpiece of a miraculous provision.

2 Kings 4:2-4 records Elisha's instructions:

"Elisha replied to her, 'How can I help you? Tell me, what do you have in your house?' 'Your servant has nothing there at all,' she said, 'except a small jar of olive oil.' Elisha said, 'Go around and ask all your neighbors for empty jars. Don't ask for just a few. Then go inside and shut the door behind you and your sons. Pour oil into all the jars, and as each is filled, put it to one side.'"

Faith in Action

The widow's faith and obedience were critical to the unfolding miracle. She and her sons went out and gathered as many empty vessels as they could from their neighbors. Once they had collected the jars, they shut the door behind them as instructed by Elisha. The act of closing the door signifies a moment of private faith, a sacred space where the miracle would take place away from public scrutiny.

The Miracle of Multiplication

The widow began to pour oil from her small jar into the empty vessels. Miraculously, the oil kept flowing. Each vessel was filled to the brim, and as one was filled, it was set aside and another brought forward. The oil only stopped flowing when there were no more vessels left to fill.

2 Kings 4:5-6 describes the miracle:

"She left him and shut the door behind her and her sons. They brought the jars to her and she kept pouring. When all the jars were full, she said to her son, 'Bring me another one.' But he replied, 'There is not a jar left.' Then the oil stopped flowing."

Divine Provision and Abundance

Elisha then instructed the widow to sell the oil, pay off her debts, and live on what remained. This miracle not only provided enough to settle her debts but also ensured that she and her sons could live off the surplus. This outcome highlights the abundance of God's provision when one steps out in faith and obedience.

2 Kings 4:7 concludes the account:

"She went and told the man of God, and he said, 'Go, sell the oil and pay your debts. You and your sons can live on what is left.'"

Lessons from the Miracle

The miracle of the widow's oil teaches several key lessons:

1. Faith and Obedience: The widow's faith in Elisha's words and her obedience in following his instructions were crucial. She acted in faith, gathering jars and pouring oil, trusting in God's provision.

2. God's Abundant Provision: The miracle illustrates that God's provision is not just sufficient but abundant. The oil continued to flow until every vessel was filled, demonstrating God's capacity to meet needs beyond immediate requirements.

3. Divine Multiplication: The small jar of oil, when placed in God's hands through faith, was multiplied to meet a significant need. This principle of divine multiplication is seen throughout the Bible, where God takes what little we have and multiplies it for His glory and our good.

4. Private Faith: The instruction to shut the door signifies that sometimes, miracles and acts of faith occur in private, away from public view. This private faith is a powerful

testimony to the personal relationship and trust one has in God.

5. Preparation and Capacity: The widow's preparation—collecting as many jars as possible—directly influenced the extent of the miracle. Her preparation in faith determined the capacity of God's blessing.

The miracle of the widow's oil is a timeless reminder of God's ability to provide in miraculous ways. It challenges believers to act in faith, prepare diligently, and trust in God's abundant provision. This story, nestled in the pages of the Old Testament, continues to inspire and assure believers of God's unwavering faithfulness and power to meet every need.

Mathematical Proof of the Miracle of the Widow's Oil

The miracle of the widow's oil in 2 Kings 4:1-7 is a powerful testament to divine provision and faith. While miracles by nature transcend human understanding, we can explore the principles behind them using mathematical concepts to appreciate the depth of these divine acts. This chapter aims to provide a mathematical perspective on the miracle, demonstrating how the multiplication of the widow's oil can be understood through the lens of mathematics.

The Story Recap

To recap, the widow was instructed by Elisha to gather empty vessels and pour her small amount of oil into them.

The oil miraculously multiplied, filling all the vessels until there were no more left to fill. The key elements of the story for our mathematical exploration are:

1. Initial Quantity: One small jar of oil

2. Action: Pouring oil into many empty vessels

3. Result: All vessels were filled, and the oil stopped only when there were no more vessels left

Understanding the Miracle Mathematically

While miracles defy natural laws and cannot be fully explained by human logic, we can use mathematical concepts to illustrate the nature of the miracle.

1. Concept of Infinite Supply

Mathematically, the miracle can be viewed through the concept of an infinite supply. If we consider the small jar of oil as having an infinite potential to be divided, the process of pouring oil into the vessels can be represented as:

- Let O be the initial amount of oil in the jar.

- Let V be the number of empty vessels.

The miracle implies that for each vessel v_i_where (i) ranges from 1 to V, there was always enough oil to fill v_i from O.

2. Division and Distribution

In mathematical terms, if the initial oil O has infinite divisibility, we can represent the filling of each vessel as a fraction of O:

$$O \rightarrow \sum_{i=1}^{v} \quad O_i$$

where O_i is the amount of oil in the i-th vessel.

Since the oil did not run out until the last vessel was filled, it implies:

$$O \rightarrow \sum_{i=1}^{v} \quad \leq O_i$$

This suggests that O was effectively infinite in its ability to supply oil as long as there were vessels to fill.

3. Infinite Series Concept

To further understand this, we can use the concept of an infinite series. In mathematics, an infinite series is a sum of infinitely many terms. If we consider the small jar of oil as an infinite series, each term of the series represents the amount of oil poured into each vessel:

$$O = O_1 + O_2 + O_3 \dots + O_v$$

Since the oil kept flowing until the vessels were filled, the series converges to the initial amount O:

$$O = \sum_{i=1}^{v} \quad O_i$$

4. Faith and Obedience as Variables

In this miracle, faith and obedience play critical roles. Mathematically, we can consider faith and obedience as variables that influence the outcome. Let F represent faith and

O_b represent obedience. The multiplication of oil can be expressed as a function of these variables:

$$M = f(O, F, O_b)$$

where M is the miraculous multiplication.

If we assume F and O_b are at their maximum (infinite faith and perfect obedience), the function f results in an infinite multiplication of oil:

$$M = \infty$$

5. Finite Perspective

From a finite perspective, let's consider the scenario where the number of vessels V is finite. The miracle's impact can be illustrated using the concept of proportions. If the initial oil O was divided among V vessels such that each vessel received a sufficient amount:

$$O_i = O/V$$

However, since the oil did not diminish and each vessel was filled to the brim, the effective oil O becomes:

$$O = V \text{ times } O_i$$

Given that O was a small jar, the multiplication implies a miraculous expansion beyond natural means.

Mathematically, the miracle of the widow's oil showcases the principle of divine multiplication, where a limited resource becomes effectively infinite through divine intervention. While human logic and mathematics can illustrate the process, they cannot fully encapsulate the divine nature of miracles. This exploration using mathematical concepts like infinite supply, series, and proportional distribution helps us appreciate the magnitude of the miracle and the power of faith and obedience in unlocking divine provision.

The story of the widow's oil remains a testament to God's ability to provide abundantly, turning scarcity into surplus through the miraculous multiplication of resources.

CHAPTER 05

GIDEON'S ARMY REDUCTION (JUDGES 7:1-8)

The story of Gideon's army reduction in Judges 7:1-8 is a remarkable demonstration of God's power and a testament to the principle that victory is achieved not by human strength but by divine intervention. Gideon's journey from leading a large force to a small, seemingly insignificant band of warriors underscores the importance of faith and reliance on God. This chapter explores the reduction of Gideon's army, the reasons behind it, and the profound lessons it teaches about trust in divine strength.

The Story Recap

Gideon, chosen by God to deliver Israel from the oppression of the Midianites, started with an army of 32,000 men. However, God had a different plan. He intended to show His power and ensure that Israel recognized that their victory came from Him alone. Through a series of reductions, Gideon's army was whittled down to a mere 300 soldiers.

Initial Quantity: 32,000 Soldiers

The initial number of soldiers in Gideon's army was 32,000. This was a sizable force, yet God saw it necessary to reduce this number drastically. The reason for this was explicitly stated in Judges 7:2: "The Lord said to Gideon, 'You have too many men. I cannot deliver Midian into their hands, or Israel would boast against me, saying, 'My own strength has saved me.'"

The First Reduction: Fearful Soldiers

God's first instruction to Gideon was to announce to the army that anyone who was fearful or trembling should go home. This call to reduce the numbers resulted in 22,000 men leaving, leaving Gideon with 10,000 soldiers. The departure of the fearful soldiers illustrates a key principle: courage and faith are essential in the face of overwhelming odds.

The Second Reduction: The Water Test

Despite the reduction, God declared that the remaining 10,000 were still too many. He instructed Gideon to take the men to the water and observe how they drank. Those who lapped the water with their tongues, like a dog, were set apart from those who knelt down to drink. Only 300 men lapped the water with their hands to their mouths, and it was these 300 that God chose to deliver Israel.

Final Quantity: 300 Soldiers

The final number of soldiers in Gideon's army was 300. This minuscule force was up against a vast Midianite army, described in Judges 7:12 as "thick as locusts," with camels that could no more be counted than the sand on the seashore. From a human perspective, this seemed like a recipe for certain defeat. Yet, it was precisely through this drastic reduction that God intended to demonstrate His power and ensure that the victory could only be attributed to Him.

Result: Victory Through Divine Intervention

With just 300 men, Gideon defeated the Midianites. The strategy was divinely inspired and unconventional. Each man was equipped with a trumpet, an empty jar, and a torch inside the jar. They surrounded the Midianite camp at night, and on Gideon's signal, they blew their trumpets, smashed the jars, and held up the torches. The sudden noise, lights, and confusion caused the Midianites to panic and turn on each other. In the ensuing chaos, the Midianites fled, and Israel achieved a stunning victory.

Mathematical Perspective: Reduction and Multiplication

From a mathematical standpoint, the reduction of Gideon's army can be viewed through the principles of subtraction and multiplication:

1. Subtraction:

- Initial number: 32,000 soldiers

- First reduction: 32,000 - 22,000 = 10,000 soldiers

- Second reduction: 10,000 - 9,700 = 300 soldiers

2. Multiplication of Effectiveness:

- Despite the reduction, the effectiveness of the 300 soldiers was multiplied exponentially through God's intervention. The strategy employed made each soldier appear much more formidable than their actual number, leading to a multiplication of fear and confusion among the enemy.

Trust in Divine Strength

The reduction of Gideon's army serves as a powerful reminder that true strength and victory come from trusting in God's power rather than relying on human resources. The drastic reduction ensured that no one could boast of their own strength, highlighting the principle found in Zechariah 4:6: "Not by might nor by power, but by my Spirit,' says the Lord Almighty."

Lessons Learned

1. Faith Over Numbers:

- This story teaches us that having faith in God is more important than the number of resources we possess. God can use even a small, faithful remnant to achieve great victories.

2. Obedience to God's Instructions:

- Gideon's obedience to God's seemingly illogical instructions was crucial. It is a reminder that divine wisdom often transcends human understanding.

3. God's Glory:

- The ultimate purpose of the reduction was to ensure that the glory of the victory belonged to God alone. This teaches us to attribute our successes to God and not to our own efforts.

The reduction of Gideon's army from 32,000 to 300 soldiers and the subsequent victory over the Midianites is a profound illustration of divine strength and provision. By reducing the army to an impossibly small number, God demonstrated that victory is not achieved by human might but by His power. This chapter underscores the importance of faith, obedience, and the recognition of God's hand in our victories, reminding us that true success comes from reliance on Him.

Proving the Miracle of Gideon's Army Mathematically

The story of Gideon's army reduction in Judges 7:1-8 is a testament to God's power and a profound illustration of faith and reliance on divine strength. Mathematically analyzing this event provides insight into how divine intervention transcends human limitations. This chapter explores how the miraculous reduction of Gideon's army can be understood through mathematical principles and probability, emphasizing the improbability of victory by human standards and the certainty of divine power.

Initial Conditions and Reductions

Initial Quantity: 32,000 Soldiers

Gideon started with an army of 32,000 soldiers, facing a Midianite force described as innumerable. This large initial number represented a formidable human strength.

First Reduction: Fearful Soldiers

God instructed Gideon to announce that anyone who was afraid could leave. This resulted in 22,000 soldiers departing, leaving 10,000. The reduction can be expressed as:

Remaining soldiers = 32,000 - 22,000 = 10,000

Second Reduction: The Water Test

The second reduction involved a unique test at the water. Only those who lapped the water with their hands to their mouths were chosen. Out of 10,000, only 300 passed this test:

Remaining soldiers = 10,000 - 9,700 = 300

Probability and Divine Selection

To appreciate the improbability of victory by human standards, consider the probability of selecting 300 soldiers out of 32,000 through two rounds of reduction.

Probability of First Reduction

The first reduction left $\dfrac{10{,}000}{32{,}000}$ or 1/3.2 of the soldiers. The probability P_1 of any soldier being among the remaining 10,000 is:

$$P_1 = \dfrac{10{,}000}{32{,}000} = 10{,}000/32{,}000 = 0.3125$$

Probability of Second Reduction

The second reduction left $\dfrac{300}{10{,}000} = 300/10{,}000$ or 1/33.33 of the soldiers. The probability P_2 of any soldier being among the remaining 300 is:

$$P_2 = \dfrac{300}{10{,}000} \quad 300/10{,}000 = 0.03$$

Combined Probability

The combined probability P of a soldier making it through both reductions is the product of the individual probabilities:

$$P = P_1 \text{ times } P_2 = 0.3125 \text{ times } 0.03 = 0.009375]$$

Thus, the chance of any single soldier being among the final 300 is less than 1%.

Military Improbability

Given the vast Midianite army, the improbability of victory for a mere 300 soldiers becomes even starker. Assuming the Midianite force was indeed as numerous as described, the ratio of soldiers was likely overwhelming. For instance, if the Midianites numbered 135,000 (as estimated in Judges 8:10), the ratio was:

$$\text{Ratio} = \frac{135,000}{300} = 135,000/300 = 450$$

This means each of Gideon's soldiers faced 450 Midianites, an impossible task by human standards.

Divine Intervention

Mathematical Miracle

The improbability calculated above underscores the necessity of divine intervention. The miraculous victory is a vivid illustration that the battle was won not by human strength or numbers but by God's power. The strategy employed by Gideon's 300 men further emphasizes this point.

Strategic Multiplication

The strategy of using trumpets, jars, and torches was designed to create the illusion of a much larger force. This psychological warfare multiplied the perceived number of soldiers:

Perceived soldiers = 300 times (effect multiplier)

If each soldier's actions made it seem like there were 10 soldiers, the perceived size was:

300 times 10 = 3,000

However, the confusion and fear sown by the noise and lights caused the Midianites to turn on each other, amplifying the effectiveness of Gideon's small force.

Lessons in Faith and Mathematics

Faith Over Numbers

The reduction from 32,000 to 300 soldiers teaches that victory depends on faith in God rather than human strength. The mathematical improbability of victory by human means emphasizes the importance of trusting in divine power.

Obedience to Divine Instructions

Gideon's adherence to God's instructions, despite their seeming illogicality, resulted in an incredible victory. This highlights the importance of obedience to divine guidance, even when it defies human understanding.

Recognition of Divine Glory

The ultimate purpose of the reduction was to ensure that the victory could only be attributed to God. This principle teaches us to recognize and attribute our successes to divine intervention rather than our efforts.

The story of Gideon's army reduction, when analyzed mathematically, reveals the astonishing improbability of victory by human means and the absolute necessity of divine intervention. Through faith, obedience, and recognition of God's power, the seemingly impossible was achieved. This chapter underscores the profound lessons of trust in divine strength and the recognition of God's hand in our victories, reminding us that true success comes not from human might but from reliance on God.

CHAPTER 06

JONAH IN THE BELLY OF THE FISH: SCIENTIFIC AND MATHEMATICAL PERSPECTIVES

The story of Jonah in the belly of the fish is one of the most fascinating and debated miracles in the Bible. Jonah 1:17 states: "Now the Lord had prepared a great fish to swallow Jonah. And Jonah was in the belly of the fish three days and three nights." This period is not only significant in its own right but also serves as a foreshadowing of Jesus' resurrection. This chapter explores the mathematical significance of the three days and three nights and provides scientific perspectives on how Jonah might have survived this ordeal.

Mathematical Significance of Three Days and Three Nights

The period of three days and three nights appears multiple times in the Bible and holds significant theological weight:

1. Foreshadowing of Jesus' Resurrection:

- Jesus referred to Jonah's three days and nights in the belly of the fish as a foreshadowing of His own resurrection. In Matthew 12:40, Jesus says, "For as Jonah was three days and three nights in the belly of the great fish, so will the Son of Man be three days and three nights in the heart of the earth."

- This parallel emphasizes the importance of the number three in salvation history, symbolizing completeness and divine intervention.

2. Symbolism of the Number Three:

- In biblical numerology, the number three often signifies completeness or perfection. Examples include the Trinity (Father, Son, Holy Spirit), Jesus' resurrection on the third day, and Jonah's three days in the fish.

- The repetition of the number three highlights God's power and the fulfillment of His promises.

Scientific Perspectives on Jonah's Survival

While the Bible describes Jonah's survival as a miraculous event, some have sought to understand it through scientific perspectives. Here are several theories that attempt to explain how Jonah might have survived inside a fish for three days and nights:

1. Natural Explanation:

- Some scholars suggest that Jonah was swallowed by a species of large fish or whale capable of swallowing a human whole. The Sperm Whale and the Whale Shark are often cited as possibilities.

- Inside such an animal, Jonah might have found an air pocket that allowed him to breathe. The stomach acids could have caused significant discomfort but may not have been immediately fatal.

2. Suspended Animation:

- Another theory posits that Jonah might have entered a state of suspended animation or hibernation. In this state, the body's metabolic processes slow down dramatically, reducing the need for oxygen and allowing survival in low-oxygen environments.

3. Divine Intervention:

- From a faith perspective, many believers hold that Jonah's survival was purely a miraculous act of God. God, who created the laws of nature, is not bound by them and can intervene supernaturally to preserve life.

The Biological Possibility

1. Air Supply:

- A fish's stomach does not have a natural air supply; however, some fish have multiple stomachs, which might include a storage chamber with trapped air. If Jonah was in such a chamber, he could have had access to enough air to breathe for a short period.

2. Protection from Digestive Enzymes:

- Digestive enzymes and stomach acids in a large fish would start breaking down organic matter. Jonah would need divine protection or a natural mechanism, such as a thick mucus lining, to prevent digestion.

3. Buoyancy and Positioning:

- Jonah's position within the fish could have affected his chances of survival. If he was in a part of the stomach with less digestive activity, his chances of surviving would increase.

Historical and Cultural Context

1. Ancient Near Eastern Literature:

- The story of Jonah shares similarities with other ancient Near Eastern literature where heroes survive within sea creatures. This common motif might reflect a broader cultural understanding of divine deliverance from the sea.

2. Jesus' Reference to Jonah:

- Jesus' reference to Jonah's ordeal provides additional weight to its significance. By linking Jonah's experience to His own death and resurrection, Jesus underscores the miraculous nature of both events.

The account of Jonah in the belly of the fish for three days and three nights remains a profound and mysterious event in the Bible. Mathematically, the number three underscores themes of completeness and divine intervention, connecting Jonah's experience to the resurrection of Jesus. Scientifically, while natural explanations and theories of suspended animation provide intriguing possibilities, the miraculous nature of the event cannot be discounted.

Jonah's survival is a testament to God's power to save and deliver, serving as a powerful symbol of resurrection and hope. The story invites believers to trust in divine providence and the fulfillment of God's promises, just as Jonah was delivered from the depths and Jesus rose from the grave. Through this lens, the miracle of Jonah's three days and three nights in the belly of the fish continues to inspire faith and wonder.

CHAPTER 07

THE SEVENTY WEEKS PROPHECY (DANIEL 9:24-27)

The Seventy Weeks Prophecy found in Daniel 9:24-27 is one of the most intricate and significant prophecies in the Bible. It provides a detailed timeline for critical events in Jewish and Christian history, including the coming of the Messiah and the end times. This prophecy's numerical aspects have intrigued scholars and believers alike, highlighting God's sovereignty over time.

The Prophecy Explained

The prophecy given to Daniel by the angel Gabriel is as follows:

> "Seventy weeks are decreed upon thy people and upon thy holy city, to finish the transgression, and to make an end of sins, and to make reconciliation for iniquity, and to

bring in everlasting righteousness, and to seal up vision and prophecy, and to anoint the most Holy.

> Know therefore and understand, that from the going forth of the commandment to restore and to build Jerusalem unto the Messiah the Prince shall be seven weeks, and threescore and two weeks: the street shall be built again, and the wall, even in troublous times.

> And after threescore and two weeks shall Messiah be cut off, but not for himself: and the people of the prince that shall come shall destroy the city and the sanctuary; and the end thereof shall be with a flood, and unto the end of the war desolations are determined.

> And he shall confirm the covenant with many for one week: and in the midst of the week he shall cause the sacrifice and the oblation to cease, and for the overspreading of abominations he shall make it desolate, even until the consummation, and that determined shall be poured upon the desolate." (Daniel 9:24-27, KJV)

Understanding the Time Frame

1. Seventy Weeks: The term "weeks" in this context is generally understood to mean "weeks of years," where one week equals seven years. Therefore, seventy weeks represent a period of $70 \cdot 7 = 490$ years.

2. Divisions of the Seventy Weeks:

- 7 weeks: 7 · 7 = 49 years

- 62 weeks: 62 · 7 = 434 years

- 1 week: 1 · 7 = 7 years

Breakdown of the Prophecy

1. The Command to Restore and Rebuild Jerusalem:

- The prophecy begins with a decree to rebuild Jerusalem. This is historically linked to the decree issued by Artaxerxes I in 445 B.C. (Nehemiah 2:1-8).

2. First Division: 7 Weeks (49 Years):

- This period covers the time it took to rebuild Jerusalem, including the walls and streets, in "troublous times." This rebuilding effort, as recorded in the book of Nehemiah, was completed around 396 B.C.

3. Second Division: 62 Weeks (434 Years):

- Following the rebuilding, an additional 434 years pass until the coming of "Messiah the Prince." This timeline brings us to the period of Jesus' ministry. Calculating from 396 B.C. and adding 434 years, we arrive around 27-30 A.D., the time of Jesus' crucifixion.

4. The Final Week (7 Years):

- The last week is divided into two parts. The prophecy mentions that in the middle of this week (3.5 years), sacrifices will cease, and abominations will occur. This period is often associated with end-time events and the Antichrist, as detailed in other prophetic books like Revelation.

Mathematical Verification

1. Total Calculation:

- 7 weeks (49 years) + 62 weeks (434 years) + 1 week (7 years) = 490 years.

2. Key Historical Milestones:

- Start Point: 445 B.C. (Decree to rebuild Jerusalem)

- Rebuilding Completed: 396 B.C. (49 years later)

- Messiah's Arrival: 27-30 A.D. (434 years after 396 B.C.)

- End Time Prophecies: Future fulfillment during the last 7 years.

Mathematical Precision in Prophecy

1. From Decree to Jesus:

- The prophecy's precision in predicting the arrival of the Messiah is remarkable. From the decree to rebuild Jerusalem (445 B.C.) to the approximate time of Jesus' crucifixion (27-30 A.D.) is about 483 years, aligning with the 69 weeks (7 + 62).

2. Symbolic Significance of Numbers:

- The use of 7, a number often representing completeness and perfection in biblical literature, underscores the divine orchestration of these events.

Theological Implications

1. Divine Sovereignty:

- The exactness of this prophecy demonstrates God's control over history. The foretelling of specific events centuries before they occurred highlights His omniscience and omnipotence.

2. Messianic Fulfillment:

- Jesus' life and ministry fulfill the prophecy's timeline, reinforcing the belief in Him as the promised Messiah.

3. End-Time Events:

- The final week remains a subject of eschatological study, pointing to future events that will culminate in the ultimate fulfillment of God's plan for humanity.

The Seventy Weeks Prophecy in Daniel 9:24-27 serves as a powerful testament to God's precise control over time and history. By examining the mathematical breakdown of the prophecy, we see a clear timeline leading from the decree to rebuild Jerusalem to the ministry of Jesus Christ and the anticipated end times. This prophecy not only strengthens faith in the reliability of Scripture but also provides a framework for understanding God's redemptive plan through history. The exactitude and significance of this prophecy underscore the importance of trusting in God's sovereignty and the fulfillment of His promises.

58

CHAPTER 08

THE NUMBER OF THE BEAST (REVELATION 13:18)

The prophecy concerning the number of the beast, found in Revelation 13:18, has fascinated and puzzled scholars and believers for centuries. This chapter will explore the significance of the number 666, its mathematical implications, and the possible interpretations that reveal its prophetic and divine nature.

The Prophecy Explained

Revelation 13:18 states:

> "Here is wisdom. Let him that hath understanding count the number of the beast: for it is the number of a man; and his number is Six hundred threescore and six" (KJV).

Understanding the Number 666

1. Numerical Representation:

- The number 666 is described explicitly as "six hundred threescore and six," which translates to 600 + 60 + 6.

2. Symbolic Meaning:

- In biblical numerology, the number 6 often symbolizes imperfection or humanity, falling short of the divine number 7, which represents completeness and perfection.

Historical and Mathematical Analysis

1. Gematria:

- Gematria is an alphanumeric code where letters are assigned numerical values. This ancient practice was used in Hebrew and Greek texts to find hidden meanings.

- For example, in Greek, the name "Nero Caesar" transliterates into Hebrew as "קסר נרון" (Neron Kesar). Assigning numerical values to these letters (Nun=50, Resh=200, Vav=6, Nun=50, Qoph=100, Samekh=60, Resh=200), we get:

- נ (Nun) = 50

- ר (Resh) = 200

- ו (Vav) = 6

- ן (Nun) = 50

- ק (Qoph) = 100

- ס (Samekh) = 60

- ר (Resh) = 200

- Adding these values: 50 + 200 + 6 + 50 + 100 + 60 + 200 = 666

2. The Sum of the First Six Digits:

- Another mathematical way to interpret 666 is by summing the first six natural numbers:

- $1 + 2 + 3 + 4 + 5 + 6 = 21$

- 21 can be multiplied by 31.71428571 (21·31.71428571 ≈ 666), indicating a complex relationship among numbers, though the sum of squares method is more direct and accurate.

- The sum of squares i.e., $1^2 + 2^2 + 3^2 + 4^2 + 5^2 + 6^2$ gives $1 + 4 + 9 + 16 + 25 + 36 = 91$, and this multiplied by 7.31978 gives ≈ 666, an interesting, albeit indirect, connection.

3. Triangular Number:

- A triangular number is a number that can form an equilateral triangle. The formula to find the nth triangular number is $T_n = \dfrac{n(n+1)}{2}$

- For 666, we reverse this to find n:

- $666 = \dfrac{n(n+1)}{2}$

- Solving for n: $(n^2 + n - 1332 = 0)$

- Using the quadratic formula n $= \frac{-b \pm \sqrt{b^2 - 4ac}}{2a}$

we get $n = \frac{-1 \pm \sqrt{1 - 5328}}{2}$

- Approximating, n $= \frac{-1 \pm 73}{2}$

- Taking the positive root, n ≈ 36, indicating the significance in numerological and mathematical patterns.

Interpreting the Number 666

1. Historical Context:

- In the context of Revelation, the beast represents a worldly power opposing God. The use of 666 could indicate a specific historical figure known for persecuting Christians, such as Emperor Nero.

- The numeric value of Nero's name in Hebrew, as shown in Gematria, supports this interpretation.

2. Theological Implications:

- The recurrence of the number 6 highlights humanity's imperfection and rebellion against God's perfect order.

- Three sixes (666) emphasize complete imperfection, the antithesis of divine perfection (777).

3. Modern Speculations:

- Over the years, many have speculated about the identity of the beast, linking the number 666 to various historical and contemporary figures through the creative use of numerology.

- However, caution must be exercised to avoid unfounded or sensational interpretations.

Mathematical Proof and Divine Revelation

1. Gematria Verification:

- The calculation through Gematria is a direct way to connect historical figures to the prophecy, providing a clear mathematical correlation.

2. Triangular Number Significance:

- The triangular number approach provides another layer of understanding, showcasing the complex and intricate design of biblical prophecy.

3. Symbolic Summation:

- Summing the first six natural numbers and understanding their symbolic meaning (imperfection) aligns with the theological message of Revelation.

The number of the beast, 666, as described in Revelation 13:18, serves as a profound and multifaceted symbol within biblical prophecy. Through historical analysis, Gematria, and mathematical approaches, we can gain insights into its significance and implications. This number underscores humanity's imperfection and opposition to divine perfection, reminding believers of the ultimate victory of God's perfect plan. As we explore the depth of this prophecy, we are reminded of the sovereignty and wisdom of God in revealing His truths through both scripture and the language of mathematics.

CHAPTER 09

THE WALLS OF JERICHO (JOSHUA 6:1-27)

The Fall of the Walls of Jericho is one of the most dramatic and well-known stories in the Bible, showcasing God's power and the importance of obedience to His instructions. This chapter will explore the narrative, the specific numerical instructions involved, and the mathematical implications of this miraculous event.

The Biblical Account

In Joshua 6:1-27, the Bible describes how the Israelites, under Joshua's leadership, captured the fortified city of Jericho. The strategy for this victory was unique and divinely orchestrated.

Instructions and Actions

1. Days of Marching:

- The Israelites were instructed to march around the city once per day for six days.

2. Priests with Trumpets:

- Seven priests were to carry seven trumpets of rams' horns before the Ark of the Covenant.

3. Circuits on the Seventh Day:

- On the seventh day, they were to march around the city seven times, and then the priests were to blow the trumpets.

4. Outcome:

- When the priests made a long blast with the ram's horn and the people shouted, the walls of Jericho collapsed, and the Israelites took the city.

Mathematical and Symbolic Implications

1. The Number Seven:

- The number seven is significant in biblical numerology, often representing completeness, perfection, and divine order.

- The repeated use of the number seven in the instructions emphasizes the perfection of God's plan and the importance of following His commands precisely.

2. Geometric Progression:

- The instructions can be viewed as a geometric progression, where the actions on each day build upon the previous days, culminating in the final act on the seventh day.

- This progression highlights the importance of persistence and faith in achieving the desired outcome.

3. Obedience and Miracles:

- The precise actions taken by the Israelites, following God's instructions exactly, demonstrate a direct correlation between obedience and miraculous outcomes.

- This principle can be expressed mathematically as a function where obedience (O) leads to a miraculous outcome (M), i.e., $M = f(O)$, where (f) represents the divine function.

Mathematical Analysis

1. Seven Priests and Seven Trumpets:

- The use of seven priests with seven trumpets signifies a complete and perfect preparation for the divine intervention.

- Mathematically, this can be viewed as 7 times $7 = 49$, symbolizing a square of completeness.

2. Circuits on the Seventh Day:

- Marching around the city seven times on the seventh day signifies a doubling of the divine completeness, reinforcing the importance of this final act.

- The total number of circuits made around the city is 6 + 7 = 13. While the number 13 is often associated with rebellion or judgment in numerology, in this context, it signifies the judgment of Jericho and the completeness of God's plan.

3. Logarithmic Growth:

- The intensity of the actions increases logarithmically, with a constant (once per day) for six days and an exponential increase (seven times) on the seventh day.

- This growth can be represented mathematically by the formula $y = \log_b(x)$, where (y) is the intensity of the action, (x) is the day, and (b) is the base of the logarithm representing the divine multiplier.

Symbolic Significance

1. Perfection and Completeness:

- The number seven, used repeatedly in this story, underscores the perfection and completeness of God's plan.

- This symbolic use of numbers reminds believers of the importance of trust and obedience to God's instructions.

2. Faith and Action:

- The Israelites' faith in following what might have seemed like unusual instructions resulted in a miraculous victory.

- This principle can be applied to modern life, where faith combined with precise actions can lead to extraordinary outcomes.

3. Divine Strategy:

- The strategic use of numbers in this story shows that God's plans are not arbitrary but are meticulously designed with specific purposes.

- Understanding these numerical patterns can deepen one's appreciation of divine wisdom and order.

The fall of the walls of Jericho, as described in Joshua 6:1-27, is a powerful testament to the importance of obedience to God's instructions and the miraculous outcomes that result from such obedience. The specific numerical instructions involving the number seven highlight the completeness and perfection of God's plan. Through mathematical analysis and symbolic interpretation, this story illustrates the profound connection between faith, action, and divine intervention. Believers are reminded that by following God's precise instructions, they can experience His power and miracles in their lives, just as the Israelites did at Jericho.

CHAPTER 10

THE MIRACLE AT CANA – TURNING WATER INTO WINE

The miracle of Jesus turning water into wine at the wedding in Cana is one of the first public miracles He performed, as recorded in the Gospel of John. This event not only signifies Jesus' divine power but also carries profound symbolic meanings and implications for the Christian faith. In this chapter, we will explore the narrative, the significance of the miracle, and its theological and symbolic meanings.

The Biblical Account

The story is found in John 2:1-11:

Setting the Scene

1. The Occasion:

- The miracle took place during a wedding feast in Cana of Galilee, an event of significant social importance in Jewish culture.

2. The Problem:

- The hosts ran out of wine, which would have been a social embarrassment and a hospitality failure.

3. Mary's Intervention:

- Mary, the mother of Jesus, informed Him of the shortage, implying her belief in His ability to remedy the situation.

The Miracle

1. Jesus' Response:

- Jesus initially responded, "My hour has not yet come," indicating a future time for His full revelation.

2. Mary's Faith:

- Despite Jesus' initial response, Mary instructed the servants to do whatever Jesus told them, demonstrating her faith in His divine authority.

3. The Instructions:

- Jesus directed the servants to fill six stone water jars, used for ceremonial washing, with water. These jars held about 20 to 30 gallons each.

4. The Transformation:

- Jesus then instructed the servants to draw some out and take it to the master of the banquet. When the master tasted the water that had been turned into wine, he was astonished by its quality.

The Significance

1. First Sign:

- This miracle is noted as the first of Jesus' signs, revealing His glory and leading His disciples to believe in Him.

2. Symbolic Meaning:

- The transformation of water into wine symbolizes the new covenant and the joy and abundance that Jesus brings.

- Water, used for purification rites, being turned into wine, a symbol of celebration, signifies the fulfillment and transformation of the old covenant into the new through Jesus.

3. Quality of Wine:

- The high quality of the wine produced by Jesus signifies the superior nature of the new covenant over the old.

Theological Implications

1. Divine Authority:

- This miracle demonstrates Jesus' divine authority over nature, affirming His identity as the Son of God.

2. Provision and Abundance:

- Jesus' ability to provide abundantly in a time of need illustrates God's provision and generosity towards His people.

3. Transformation:

- The miracle serves as a metaphor for spiritual transformation, indicating that Jesus has the power to transform the ordinary into the extraordinary.

Mathematical and Symbolic Analysis

1. Quantitative Aspect:

- Six stone jars, each holding 20 to 30 gallons, means Jesus transformed between 120 and 180 gallons of water into wine. This large quantity underscores the abundance of Jesus' provision.

2. Number Six:

- The number six often represents imperfection or incompletion in biblical numerology, suggesting that Jesus' miracle brings completeness and perfection.

3. Symbolism of Wine:

- Wine is often associated with joy, celebration, and the messianic banquet in the Bible (Isaiah 25:6). By providing wine, Jesus not only met a physical need but also pointed to the eschatological joy and abundance of God's kingdom.

Faith and Obedience

1. Mary's Faith:

- Mary's confidence in Jesus' ability to address the situation exemplifies deep faith and trust in His divine mission.

2. Servants' Obedience:

- The obedience of the servants, even when the instructions seemed nonsensical, highlights the importance of trust and obedience in witnessing God's miracles.

The miracle at Cana, where Jesus turned water into wine, is a profound demonstration of His divine power, ability to transform, and provision. This event not only saved the hosts from social disgrace but also symbolically pointed to the transformative power of Jesus' ministry and the inauguration of the new covenant. Through this miracle, Jesus revealed His glory, deepening the faith of His disciples and illustrating key theological truths about His identity and mission. As believers reflect on this event, they are reminded of the importance of faith, obedience, and trust in Jesus' power to provide abundantly and transform lives.

The Scientific Perspective on Water to Wine Transformation

The miracle of Jesus turning water into wine at the wedding in Cana, as recorded in John 2:1-11, is one of the most celebrated events in the New Testament. From a theological standpoint, this miracle demonstrates Jesus' divine power. However, from a scientific perspective, the transformation of water into wine poses interesting questions about the feasibility of such a process. This chapter explores the scientific aspects of this miracle, examining the components of water and wine, and discussing the potential mechanisms required for such a transformation.

Components of Water and Wine

1. Water (H_2O):

- Water is a simple molecule consisting of two hydrogen atoms bonded to one oxygen atom. It is a clear, tasteless, and odorless liquid that is essential for all known forms of life.

2. Wine:

- Wine is a complex mixture containing water, ethanol (C_2H_5OH), various sugars, acids, phenolic compounds, and other organic molecules. The primary components of wine include:

- Water: The major component, making up about 85-90% of the wine.

- Ethanol: Produced by the fermentation of sugars by yeast, typically constituting 10-15% of the wine.

- Sugars: Glucose and fructose are present in varying amounts.

- Acids: Such as tartaric, malic, and citric acids, which contribute to the wine's taste.

- Phenolic Compounds: Including tannins and anthocyanins, which affect the wine's color and flavor.

Scientific Process of Wine Making

1. Fermentation:

- The process of making wine involves the fermentation of grape juice by yeast. Yeast converts sugars (glucose and fructose) into ethanol and carbon dioxide. This biochemical process is complex and requires specific conditions.

2. Aging and Maturation:

- After fermentation, wine undergoes aging and maturation, where it develops its flavor, aroma, and character. This involves various chemical reactions and the gradual development of complex organic compounds.

Hypothetical Mechanisms for Water-to-Wine Transformation

1. Instantaneous Fermentation:

- For water to turn into wine, a rapid and instantaneous process equivalent to natural fermentation would need to occur. This would involve the introduction and transformation of sugars and organic compounds into the water, along with the production of ethanol.

2. Atomic and Molecular Rearrangement:

- One possible scientific explanation could involve the rearrangement of atoms and molecules. For water to turn into wine, hydrogen and oxygen atoms in water molecules would need to combine with carbon atoms (not originally present in water) to form ethanol and other organic molecules. This process would require an immense amount of energy and precise molecular control.

3. Synthetic Creation:

- Another hypothetical method might involve synthesizing wine from basic chemical elements. This would require advanced knowledge of chemistry and biochemistry to recreate the complex mixture of compounds found in wine.

Scientific Challenges

1. Complexity of Wine:

- The complexity of wine, with its multitude of compounds, poses a significant challenge. Replicating the exact composition and balance of these compounds instantaneously is beyond current scientific capability.

2. Energy Requirements:

- The transformation of water into wine would require a substantial amount of energy to break and form chemical bonds, far exceeding any natural or artificial process known to science.

3. Natural Limitations:

- Natural fermentation and aging processes take time, specific conditions, and biological agents (yeast). Achieving this instantaneously defies known natural laws.

Theological Interpretation

From a theological perspective, the miracle at Cana is not meant to be explained by scientific means. Instead, it serves as a sign of Jesus' divine authority and power. The instantaneous transformation of water into wine demonstrates His control over natural laws, emphasizing His identity as the Son of God.

The scientific explanation for the transformation of water into wine, as described in the miracle at Cana, remains elusive and beyond the reach of current scientific understanding. While the natural process of winemaking involves fermentation and complex biochemical reactions, replicating this instantaneously requires a level of control and energy that defies natural laws. From a faith perspective, the miracle underscores the divine nature of Jesus, inviting believers to trust in His power and provision. Thus, the miracle at Cana continues to inspire awe and faith, illustrating the intersection of the natural and the supernatural in the narrative of the Bible.

CHAPTER 11

EARTH'S FREE FLOAT IN SPACE

In the book of Job, written over 3,500 years ago, we find a remarkable statement about the Earth's position in the cosmos. Job 26:7 declares, "He stretches out the north over space; He hangs the earth on nothing." This verse penned long before the advent of modern astronomy, offers a profound insight into the Earth's relationship with the vastness of space.

At a time when many ancient cultures believed that the Earth was supported by a large animal or rested on pillars, the Bible presents a radically different view. It describes the Earth as hanging freely in space, suspended by nothing. This concept is remarkably accurate, aligning with our modern understanding of the Earth's position in the solar system and the universe.

The idea of Earth's free float in space was not widely accepted or understood until the advent of modern astronomy. Today, we know that the Earth orbits the Sun in a vast expanse of space, held in its trajectory by the force of gravity. The fact that the Bible described this concept millennia ago is a testament to its divine inspiration and foresight.

The Earth's free float in space has profound implications for our understanding of the universe and our place in it. It reminds us of the incredible complexity and beauty of the cosmos and the intricate design that governs the movements of celestial bodies. It also speaks to the power and wisdom of the Creator, who set the Earth in its place and sustains it in its orbit.

As we ponder the Earth's free float in space, let us marvel at the wonders of the universe and the greatness of the One who created it all. Let us stand in awe of the God who hangs the Earth on nothing, yet sustains it with His mighty hand. And let us remember that we are but small inhabitants of a vast cosmos, yet deeply loved and cared for by the Creator of it all.

Biblical Evidence: Psalm 104:5

Psalm 104:5 in the King James Bible states, "Who laid the foundations of the earth, that it should not be removed forever." This verse, like Job 26:7, underscores the stability and orderliness of the Earth as part of God's creation. Let's delve into a comprehensive study and commentary on this verse, supported by additional biblical references, interpretation, and concordance insights.

King James Bible's References:

- Psalm 119:90: "Thy faithfulness is unto all generations: thou hast established the earth, and it abideth."

- Isaiah 45:18: "For thus saith the LORD that created the heavens; God himself that formed the earth and made it; he hath established it, he created it not in vain, he formed it to be inhabited: I am the LORD, and there is none else."

- Hebrews 1:10: "And, Thou, Lord, in the beginning, hast laid the foundation of the earth, and the heavens are the works of thine hands."

Interpretation:

Psalm 104 is a hymn of praise, celebrating God's creation and the order He has established within it. Verse 5 specifically highlights God's act of laying the foundations of the Earth, ensuring its permanence and stability. The psalmist marvels at the Creator's power and wisdom in establishing the Earth so that it cannot be moved.

The phrase "laid the foundations of the earth" suggests a deliberate and powerful act of creation. It reflects an understanding that the Earth is not a random occurrence but a result of God's intentional design. The stability of the Earth, implied by "that it should not be removed for ever," speaks to God's ongoing sustenance of His creation.

Commentary:

In ancient times, many cultures had myths about the Earth being held up by various means, such as by giants or animals. The biblical view, however, presents a different picture. It portrays God as the supreme architect who not only created the Earth but also maintains its stability.

The foundation metaphor is significant because it conveys strength, permanence, and reliability. In architecture, a foundation is the essential support that ensures the stability of the entire structure. Similarly, God's establishment of the Earth is described as a foundation that cannot be shaken, emphasizing His omnipotence and the reliability of His creation.

Furthermore, this verse aligns with the broader biblical narrative that emphasizes God's sovereign control over creation. For instance, in Psalm 119:90, the psalmist acknowledges God's faithfulness across generations and His role in establishing the Earth. Isaiah 45:18 emphasizes that God created the Earth with purpose, intending it to be inhabited, further highlighting His intentional and sustained involvement in creation.

Hebrews 1:10, quoting from the Old Testament, reaffirms the foundational role of God in creation, portraying the heavens and the Earth as the work of His hands. This continuity between the Old and New Testaments underscores a consistent biblical theme: God as the unchanging and faithful Creator.

Concordance Insights:

- Foundations (Strong's H3245 - יָסַד - yacad): To lay a foundation, establish, fix.

- Removed (Strong's H4131 - מוֹט - mot): To totter, shake, slip.

- For ever (Strong's H5769 - עוֹלָם - olam): Everlasting, perpetual, eternal.

The Hebrew word "yacad" used for "laid the foundations" implies a foundational act of establishing something firmly. "Mot" indicates instability or shaking, which the verse negates regarding the Earth. "Olam" reinforces the eternal aspect of God's creation, indicating that the Earth is established to last perpetually.

Psalm 104:5, within the broader context of Scripture, celebrates God's role as the Creator and Sustainer of the Earth. It presents a worldview that sees the Earth not as a fragile entity but as a robust creation, firmly established by God. This understanding not only inspires awe and worship but also provides a sense of security, knowing that the same God who created the Earth maintains it with His powerful and unwavering hand.

The biblical evidence from Psalms, Isaiah, and Hebrews collectively portrays a consistent and profound view of God's creation. It affirms that the Earth, hanging freely in space as described in Job, is securely founded by God's design and sustained by His power. This theological foundation encourages believers to trust in the Creator's ongoing care for the world and all its inhabitants.

CHAPTER 12

THE EARTH IS ROUND

Throughout history, humanity's understanding of the earth has evolved significantly. In ancient times, many cultures believed that the earth was flat, a concept deeply ingrained in their worldviews and mythologies. However, long before the advent of modern science, the Bible presented a different perspective on the shape of the earth. Isaiah 40:22, written approximately 2,800 years ago, states: "It is He who sits above the circle of the earth." This ancient scripture hints at the spherical nature of our planet, a fact that would later be confirmed by scientific exploration and discovery.

Biblical Reference: Isaiah 40:22

The verse in Isaiah reads, "It is He who sits above the circle of the earth, and its inhabitants are like grasshoppers; who stretches out the heavens like a curtain, and spreads them out like a tent to dwell in." The word "circle" in this context is translated from the Hebrew word "chug," which can also mean a sphere or roundness. This passage highlights the grandeur of God's creation and His supreme position above it, metaphorically placing Him above the earth's curvature.

Historical Beliefs and the Flat Earth

For much of human history, the belief in a flat earth was widespread. Ancient civilizations, including the Greeks and the Mesopotamians, had various cosmological views, many of which depicted the earth as a flat disc floating on water. These views were not just based on ignorance but were deeply embedded in the cultural and religious contexts of the times.

However, certain Greek philosophers such as Pythagoras and later Aristotle proposed a spherical earth based on observations and logical reasoning. Aristotle's arguments for a spherical earth, derived from the shape of the earth's shadow on the moon during a lunar eclipse and the changing position of the stars as one traveled north or south, laid the groundwork for future scientific inquiry.

The Bible and Early Christian Thought

The Bible, with its reference to the "circle of the earth," provided an early hint at the earth's roundness. This was significant, given that the prevailing belief at the time was that of a flat earth. Early Christian thinkers, influenced by both scripture and Greek philosophy, often embraced the idea of a spherical earth. This acceptance was not uniform, and debates persisted, but the biblical reference played a role in shaping the views of some scholars.

Christopher Columbus and Biblical Inspiration

One of the most famous historical figures influenced by biblical scripture was Christopher Columbus. In his diary, Columbus attributed his inspiration for sailing west to reach Asia to the Holy Scriptures. He wrote, "It was the Lord who put it into my mind... There is no question the inspiration was from the Holy Spirit because He comforted me with rays of marvelous illumination from the Holy Scriptures..." Columbus believed that God had guided him through the Bible, and this divine guidance played a crucial role in his historic voyage.

Columbus's belief in a round earth, inspired by scripture and supported by the growing body of geographical knowledge, led him to embark on his journey across the Atlantic. His successful voyage and the subsequent discovery of the Americas were pivotal in transforming the medieval worldview and ushering in a new era of exploration and understanding.

The Intersection of Faith and Science

The acknowledgment of the earth's roundness in the Bible and the subsequent confirmation by scientific discovery illustrate the harmonious relationship that can exist between faith and science. While the Bible is not a scientific textbook, its passages can and have provided insights that align with scientific understanding.

The journey from ancient beliefs in a flat earth to the modern acceptance of a spherical earth is a testament to humanity's quest for knowledge and truth. It shows how scripture can inspire and guide exploration and discovery, leading to a deeper understanding of the natural world.

Conclusion

Isaiah 40:22 stands as a remarkable testament to the Bible's insight into the nature of the earth long before scientific evidence emerged. The verse's reference to the "circle of the earth" predated scientific proofs and influenced thinkers like Christopher Columbus to challenge the flat earth paradigm. This chapter underscores the profound impact of biblical scripture on the pursuit of knowledge and the intersection of faith and science.

The earth is indeed round, and this understanding, inspired by scripture and confirmed by science, reflects the incredible journey of human discovery. It reminds us of the vastness of God's creation and our continuous quest to comprehend it.

Isaiah 40:22 - Expository Study and Comprehensive Commentary

Verse (King James Version):

It is he that sitteth upon the circle of the earth, and the inhabitants thereof are as grasshoppers; that stretcheth out the heavens as a curtain, and spreadeth them out as a tent to dwell in.

1. Biblical References:

1. Job 9:8 - "Which alone spreadeth out the heavens, and treadeth upon the waves of the sea."

2. Psalm 104:2 - "Who coverest thyself with light as with a garment: who stretchest out the heavens like a curtain."

3. Isaiah 42:5 - "Thus saith God the Lord, he that created the heavens, and stretched them out; he that spread forth the earth, and that which cometh out of it; he that giveth breath unto the people upon it, and spirit to them that walk therein."

4. Isaiah 44:24 - "Thus saith the Lord, thy redeemer, and he that formed thee from the womb, I am the Lord that maketh all things; that stretcheth forth the heavens alone; that spreadeth abroad the earth by myself;"

5. Isaiah 45:12 - "I have made the earth, and created man upon it: I, even my hands, have stretched out the heavens, and all their host have I commanded."

6. Isaiah 48:13 - "Mine hand also hath laid the foundation of the earth, and my right hand hath spanned the heavens: when I call unto them, they stand up together."

2. Interpretation:

1. Divine Sovereignty:

The verse emphasizes God's supreme authority and sovereignty over all creation. By stating that God sits "upon the circle of the earth," it depicts Him as a ruler over the entire globe, indicating His omnipresence and omnipotence.

2. Human Insignificance:

The comparison of the inhabitants of the earth to grasshoppers underscores human smallness and insignificance in contrast to God's grandeur and majesty. This imagery serves to humble humanity and highlight God's vastness and superiority.

3. Cosmic Imagery:

The act of God stretching out the heavens "as a curtain" and spreading them out "as a tent to dwell in" uses familiar imagery to convey the ease with which God created and sustains the universe. It portrays the heavens as a vast, expansive canopy, illustrating God's creative power and His intimate involvement with His creation.

4. The Shape of the Earth:

The phrase "circle of the earth" has often been interpreted to suggest the spherical shape of the earth, indicating a biblical acknowledgment of this concept long before it was scientifically established. This interpretation aligns with the understanding that the Bible contains insights into the natural world that align with scientific discoveries.

3. Commentary:

Isaiah 40:22 is a profound verse that offers a glimpse into the nature of God's relationship with His creation. It is part of a broader passage (Isaiah 40:12-31) that extols the greatness of God and His incomparable power and wisdom.

Divine Perspective:

God's perspective is elevated and all-encompassing. Sitting above the circle of the earth, He sees and governs all. This positioning signifies authority and control, reinforcing the belief in God's omnipotence.

Human Perspective:

From the human perspective, we are like grasshoppers, small and seemingly insignificant in the grand scheme of things. This imagery is intended to evoke humility and a sense of awe in the presence of God's greatness.

Creation Imagery:

The heavens stretched out like a curtain and spread like a tent reflects God's role as the creator and sustainer of the universe. It evokes the image of God as a master craftsman, skillfully constructing the cosmos with care and precision. This cosmic tent not only provides a dwelling place for humanity but also illustrates the order and structure of the universe.

Implications for Believers:

For believers, this verse is a call to recognize and revere God's majesty. It encourages trust in His providence, acknowledging that the Creator who fashioned the heavens and the earth is more than capable of caring for His creation. It also serves as a reminder of our place within the created order, fostering a sense of wonder and dependence on God.

4. Concordance:

Circle (חוג, chûg): This Hebrew word can mean circle, sphere, or compass, indicating roundness or a curved boundary.

Sitteth (יָשַׁב, yāšab): This verb means to sit, dwell, remain, or inhabit, often used to denote a position of authority or rest.

Stretch (נָטָה, nāṭâ): This verb means to stretch out, extend, or spread out, often used in the context of creation and God's sovereign action.

Heavens (שָׁמַיִם, šāmayim): This term refers to the sky, the visible heavens, or the abode of the stars and celestial bodies, often denoting the entire expanse of the universe.

Curtain (יְרִיעָ, yārîaʿ): A tent curtain, screen, or covering, used metaphorically to describe the heavens as a vast, stretched-out canopy.

Tent (אֹהֶל, ʾōhel): A tent or dwelling place, symbolizing the temporary yet expansive nature of the heavens spread out by God.

Isaiah 40:22 is a powerful testament to God's majesty, sovereignty, and intimate involvement with His creation. It challenges us to recognize our own smallness in the face of God's greatness and to trust in His all-encompassing care. The imagery used in this verse, from the circle of the earth to the stretched-out heavens, invites us to ponder the vastness of the cosmos and the incredible power of the Creator who fashioned it all.

CHAPTER 13

THE FIRST LAW OF THERMODYNAMICS

The relationship between scientific principles and biblical narratives has long been a subject of contemplation and discussion. One such intersection lies in the First Law of Thermodynamics and its resonance with the biblical account of creation, specifically in Genesis 2:1. This chapter explores the First Law of Thermodynamics in light of the creation account, illustrating the harmony between science and Scripture.

The First Law of Thermodynamics Explained

The First Law of Thermodynamics, also known as the Law of Conservation of Energy, asserts that energy cannot be created or destroyed in an isolated system. Instead, energy can only change forms or be transferred from one part of the system to another. This principle is foundational in physics and underpins much of our understanding of energy interactions in the universe.

In a broader sense, the law also applies to matter, stating that the total quantity of matter and energy in the universe remains constant. This unchanging nature of energy and matter is a profound concept that provides a framework for understanding the physical world.

Biblical Correlation: Genesis 2:1

Genesis 2:1 states: "Thus the heavens and the earth, and all the host of them, were finished."

The Hebrew word used for "finished" in this verse is in the past definite tense, indicating an action that was completed in the past and will not occur again. The creation, as described in the Bible, was a unique event that concluded with a state of completion. The phrase "were finished" emphasizes the finality and completeness of this creative act.

Creation and the First Law of Thermodynamics

The correlation between Genesis 2:1 and the First Law of Thermodynamics is striking. The Bible's assertion that creation was finished aligns perfectly with the scientific principle that neither matter nor energy can be created or destroyed in today's world. Both the Bible and the First Law of Thermodynamics affirm that there is no ongoing creation; it is a completed process.

Scientific and Theological Implications

1. Energy Conservation: The First Law of Thermodynamics assures us that the energy in the universe remains constant. This scientific fact can be seen as a reflection of the biblical assertion that creation was finished. Just as the universe's energy is conserved, the creative acts described in Genesis have a definitive end, highlighting the completeness and sufficiency of God's work.

2. Immutable Laws: The consistency of natural laws, such as the First Law of Thermodynamics, underscores the orderliness and reliability of God's creation. These laws, established by the Creator, govern the universe with precision and predictability, reflecting the nature of a purposeful and orderly God.

3. End of Creation: The biblical narrative indicates that after the six days of creation, God ceased His creative work. This cessation parallels the scientific understanding that no new matter or energy is being created in the universe. The creation was a unique event, once completed, it necessitated no further acts of creation.

4. Stewardship of Creation: Understanding that the amount of energy and matter is finite emphasizes the

importance of stewardship. Humans, as part of creation, have a responsibility to manage and care for the resources within this closed system. This perspective aligns with the biblical mandate for humans to tend and keep the Earth (Genesis 2:15).

Harmonizing Science and Faith

The intersection of the First Law of Thermodynamics and Genesis 2:1 is a testament to the harmony that can exist between science and faith. Far from being contradictory, scientific principles can often illuminate the truths found in Scripture. The completion of creation as described in Genesis complements the scientific understanding that the universe operates under unchanging physical laws.

The First Law of Thermodynamics and the biblical account of creation in Genesis 2:1 both affirm that the act of creation is finished. The law of conservation of energy states that matter and energy are not being created or destroyed, mirroring the biblical statement that the heavens and the earth were finished. This harmonious relationship between science and Scripture enriches our understanding of the universe and highlights the wisdom of the Creator.

As we explore the wonders of the physical world through science, we find that it consistently points back to the foundational truths laid out in the Bible. The First Law of

Thermodynamics not only strengthens our comprehension of physical phenomena but also deepens our appreciation for the divine order established at the beginning of time. Thus, science and faith together paint a coherent and majestic picture of the universe, encouraging us to marvel at the intricacy and completeness of God's creation.

Genesis 2:1 - Expository Study and Comprehensive Commentary

Verse

"Thus the heavens and the earth were finished, and all the host of them." - Genesis 2:1 (KJV)

References and Context

1. Genesis 1:1 - "In the beginning, God created the heaven and the earth."

2. Genesis 1:31 - "And God saw everything that he had made, and, behold, it was very good. And the evening and the morning were the sixth day."

3. Exodus 20:11 - "For in six days the LORD made heaven and earth, the sea, and all that in them is, and rested the seventh day: wherefore the LORD blessed the sabbath day, and hallowed it."

Interpretation

Genesis 2:1 marks the conclusion of the creation narrative that spans the first chapter of Genesis. This verse emphasizes the completion of the heavens and the earth along with all their components. The use of "finished" signifies that the creative acts were brought to their intended completion, indicating a state of fullness and perfection in creation.

The term "host of them" refers to all entities within the heavens and the earth, including celestial bodies, plants, animals, and humanity. This phrase underscores the comprehensive nature of God's creation, encompassing all living and non-living things.

Commentary

1. Completion of Creation:

- The term "finished" (Hebrew: כָּלָה, kalah) implies a definitive end to the creative process. It conveys the idea that God's creation was a complete, perfect act, requiring no further additions.

- This completion aligns with the First Law of Thermodynamics, which states that matter and energy in the universe are constant, reflecting the idea that no new creation is occurring in the present.

2. Heavens and Earth:

- "Heavens" refers to the sky, outer space, and all celestial bodies. "Earth" refers to the planet and its physical features, including land, seas, and all forms of life.

- The plural form "heavens" indicates the vast expanse of the universe, with multiple layers or dimensions, which is a common biblical perspective.

3. All the Host of Them:

- The phrase "host of them" includes all created entities, emphasizing the diversity and complexity of God's creation.

- This includes the stars, planets, and all forms of life, showcasing the intricate and ordered design of the Creator.

4. Theological Implications:

- The verse highlights God's sovereignty and omnipotence. The completeness of creation reflects God's ultimate authority and power over the universe.

- It sets the stage for the introduction of the Sabbath in the following verses, emphasizing rest and reflection on God's completed work.

5. Rest as a Divine Principle:

- The completion of creation naturally leads into the concept of rest, which is introduced in Genesis 2:2-3. This rest is not due to fatigue but is a cessation from work to enjoy and celebrate the finished creation.

- It establishes a pattern for humanity, signifying the importance of rest and reflection in our lives.

Concordance

1. Finished (כָּלָה, kalah):

- Strong's H3615: To accomplish, cease, consume, determine, end, fail, finish.

- Used in various contexts to denote completion or bringing something to an end.

2. Heavens (שָׁמַיִם, shamayim):

- Strong's H8064: The sky, visible arch where the clouds move, as well as the higher ether where the celestial bodies revolve.

- Used throughout the Old Testament to refer to the physical sky and the dwelling place of God.

3. Earth (אֶרֶץ, erets):

- Strong's H776: The earth, land, country, ground.

- Refers to the physical land and is often used to denote the entire planet.

4. Host (צָבָא, tsaba'):

- Strong's H6635: A mass of persons, figuratively, a campaign, literally or figuratively, a host, an army.

- Often used to describe the heavenly bodies (stars, planets) and angelic beings, indicating an organized and ordered creation.

Conclusion

Genesis 2:1 encapsulates the completion of God's creative work, emphasizing the thoroughness and perfection of the created universe. It reflects the divine principle of a finished creation, setting the stage for the introduction of rest and sanctification. This verse serves as a foundational statement of God's sovereignty and the ordered nature of the universe, inviting believers to reflect on the majesty and completeness of God's creation.

CHAPTER 14

SECOND LAW OF THERMODYNAMICS

Psalm 102:25-26: "Of old You founded the earth, and the heavens are the work of Your hands. Even they will perish, but You endure; and all of them will wear out like a garment" (NASB).

The Bible tells us three times that the earth is wearing out like a garment. This is what the Second Law of Thermodynamics (the Law of Increasing Entropy) states: that in all physical processes, every ordered system over time tends to become more disordered. Everything is running down and wearing out as energy is becoming less and less available for use. That means the universe will eventually "wear out"—something that wasn't discovered by science until fairly recently.

Understanding the Second Law of Thermodynamics

The Second Law of Thermodynamics is a fundamental principle of physics that describes the natural progression of energy and matter in our universe. It asserts that in any isolated system, the total entropy, or degree of disorder, will always increase over time. This law is crucial for understanding the natural tendency towards chaos and degradation in all physical processes.

Entropy can be thought of as a measure of randomness or disorder within a system. The Second Law implies that energy spontaneously spreads from regions of higher concentration to regions of lower concentration, and as it does so, it becomes less useful for doing work. This gradual process of energy dispersion and increase in entropy is why perpetual motion machines are impossible; there will always be energy losses due to inefficiency and friction.

Biblical Insights on Entropy

The notion that the universe is wearing out is not a novel concept exclusive to modern science. Psalm 102:25-26 vividly describes the earth and heavens as creations of God that are destined to "wear out like a garment." This poetic yet profound statement aligns remarkably well with the scientific understanding of entropy and the Second Law of Thermodynamics.

Isaiah 51:6 also echoes this sentiment: "Lift up your eyes to the sky, Then look to the earth beneath; For the sky will vanish like smoke, And the earth will wear out like a garment And its inhabitants will die in like manner, But My salvation will be forever, And My righteousness will not wane." This verse emphasizes the transient nature of the physical world, contrasting it with the eternal nature of God's salvation and righteousness.

Hebrews 1:10-12 reinforces this message: "And, 'You, Lord, in the beginning, laid the foundation of the earth, And the heavens are the works of Your hands; They will perish, but You remain; And they all will become old like a garment, And like a mantle, You will roll them up; Like a garment, they will also be changed. But You are the same, And Your years will not come to an end.'" Here, the author of Hebrews reiterates the temporality of the created order and the unchanging, everlasting nature of God.

The Scientific Discovery of Entropy

The concept of entropy and the Second Law of Thermodynamics were formalized in the 19th century by scientists such as Rudolf Clausius and Lord Kelvin. Clausius coined the term "entropy" in 1865, while Kelvin formulated the Second Law, stating that "no process is possible in which the sole result is the absorption of heat from a reservoir and its complete conversion into work."

Before these discoveries, the prevailing belief was that the universe operated in a more static and perpetual state. The realization that all systems tend towards disorder and that energy becomes less available for work over time was groundbreaking. This understanding led to profound insights into the nature of physical processes, heat engines, and even the ultimate fate of the universe.

Implications for the Universe

The Second Law of Thermodynamics implies that the universe is on a one-way journey towards increasing entropy and disorder. This has several significant implications:

1. Energy Degradation: As energy spreads out and becomes less concentrated, it becomes less available for doing useful work. This degradation means that over time, energy sources like stars will exhaust their fuel, and the universe will cool down.

2. Heat Death of the Universe: The ultimate consequence of increasing entropy is the so-called "heat death" of the universe, where all energy is evenly distributed, and no thermodynamic work can occur. This state would be one of maximum entropy and minimal useful energy.

3. Temporal Nature of the Physical World: The Second Law reinforces the idea that the physical universe is not eternal. It had a beginning, and it will have an end. This aligns with the biblical perspective that the created order is temporary and will eventually "wear out."

Theological Reflections

The biblical foresight regarding the entropy and eventual wearing out of the universe underscores the timeless nature of God's Word. Long before the advent of modern science, Scripture revealed the impermanent nature of the physical world, pointing towards the Creator's eternal and unchanging character.

The recognition of the universe's entropy also invites us to reflect on our place within this temporal world. While the material universe is subject to decay, God's promises and His kingdom are everlasting. This contrast calls believers to place their hope and trust not in the transient things of this world but in the eternal nature of God and His redemptive work through Jesus Christ.

2 Corinthians 4:18 reminds us: "So we fix our eyes not on what is seen, but on what is unseen, since what is seen is temporary, but what is unseen is eternal." This verse encourages a focus on the eternal truths of God's kingdom,

which stand in stark contrast to the decaying nature of the physical universe.

Conclusion

The Second Law of Thermodynamics, with its profound implications for the fate of the universe, finds a remarkable parallel in the biblical assertion that the earth and heavens will wear out like a garment. This scientific principle highlights the transient nature of all created things and the inevitable increase of disorder and decay. Yet, amidst this inevitable entropy, the Bible points us to the enduring nature of God, whose promises and righteousness remain unchanging and eternal. As we navigate a world governed by entropy, we are called to anchor our faith in the everlasting Creator, whose kingdom and glory transcend the physical universe.

Psalm 102:25-26 Expository Study and Comprehensive Commentary

Psalm 102:25-26 (KJV)

25 Of old hast thou laid the foundation of the earth: and the heavens are the work of thy hands.

26 They shall perish, but thou shalt endure: yea, all of them shall wax old like a garment; as a vesture shalt thou change them, and they shall be changed:

Context and Background

Psalm 102 is a penitential psalm, often categorized as a prayer of the afflicted, when he is overwhelmed, and pours out his complaint before the LORD. The psalmist is in deep distress, facing adversity and the frailty of life, but amidst his lament, he also acknowledges God's eternal nature and unchanging character.

Verse 25

"Of old hast thou laid the foundation of the earth: and the heavens are the work of thy hands."

References:

- Genesis 1:1 - "In the beginning God created the heaven and the earth."

- Hebrews 1:10 - "And, Thou, Lord, in the beginning hast laid the foundation of the earth; and the heavens are the works of thine hands."

Interpretation and Commentary:

This verse speaks to the eternal power and creative work of God. The psalmist acknowledges God's role as the Creator who established the earth and the heavens. By referencing the foundation of the earth and the work of God's hands, the psalmist affirms the intentional and powerful act of creation.

- Foundation of the Earth: The phrase "laid the foundation of the earth" implies stability and a firm establishment. It signifies the beginning of the created order by God's sovereign act.

- Heavens as God's Work: The heavens, often representing the universe or the sky, are depicted as crafted by God's hands, emphasizing His intricate and purposeful design.

This verse echoes other scriptural affirmations of God as the Creator, reinforcing His omnipotence and the dependability of His creation.

Verse 26

"They shall perish, but thou shalt endure: yea, all of them shall wax old like a garment; as a vesture shalt thou change them, and they shall be changed:"

References:

- Isaiah 51:6 - "Lift up your eyes to the heavens, and look upon the earth beneath: for the heavens shall vanish away like smoke, and the earth shall wax old like a garment, and they that dwell therein shall die in like manner: but my salvation shall be for ever, and my righteousness shall not be abolished."

- Hebrews 1:11-12 - "They shall perish; but thou remainest; and they all shall wax old as doth a garment; And as a vesture shalt thou fold them up, and they shall be changed: but thou art the same, and thy years shall not fail."

Interpretation and Commentary:

This verse contrasts the temporary nature of the created order with the eternal nature of God.

- Perishable Creation: The psalmist declares that the earth and the heavens, despite their grandeur, are destined to perish. This points to the transient nature of the physical universe.

- God's Endurance: In stark contrast, God is described as enduring forever. This highlights His immutability and eternal existence.

- Waxing Old Like a Garment: The imagery of creation "waxing old like a garment" vividly illustrates the aging and decaying process. Just as clothing wears out over time, so will the heavens and the earth.

- Changing Vesture: The metaphor of changing garments indicates that God has the power to transform or renew creation. This can be seen as a reference to the eschatological renewal of the heavens and the earth, as described in Revelation 21:1, where a new heaven and a new earth are established.

Theological Implications

1. God as Creator and Sustainer: These verses reinforce the belief in God as the Creator and Sustainer of the

universe. His creation, though magnificent, is not eternal in its current form.

2. Eternal Nature of God: The contrast between the perishable creation and the enduring nature of God emphasizes His eternal, unchanging character. This provides a source of comfort and hope for believers, knowing that while the world changes and decays, God remains the same.

3. Eschatological Hope: The reference to the changing of creation like a garment hints at the future transformation of the world. This aligns with the Christian hope of a new creation, where God will renew all things.

Concordance Insights

- Foundation (H3245 - יסד - yâsad): To establish, find, or appoint.

- Work (H4639 - מַעֲשֶׂה - ma'ăśeh): Deed, work, or act.

- Endure (H5975 - עמד - ʿāmad): To stand, remain, or endure.

- Wax Old (H1086 - בָּלָה - bālāh): To wear out, decay.

- Vesture (H3830 - לְבוּשׁ - lĕbûš): Garment, clothing.

These terms highlight the foundational work of God in creation and the inevitable aging of that creation, juxtaposed with God's enduring nature.

Conclusion

Psalm 102:25-26 beautifully juxtaposes the transient nature of the physical universe with the eternal and unchanging character of God. These verses remind us of God's sovereign power in creation and His constancy amidst the changing and perishing world. They encourage believers to place their trust in the everlasting Creator, whose purposes and promises endure forever, offering a hopeful perspective on the future transformation and renewal of all things.

CHAPTER 15

THE HYDROLOGIC CYCLE

The Bible is not just a spiritual guide but also a source of profound wisdom that aligns with scientific truths discovered centuries later. One such example is the hydrologic cycle, which is eloquently referenced in the Scriptures. Approximately 2,800 years ago, the prophet Amos wrote, "He…calls for the waters of the sea, and pours them out on the face of the earth…" (Amos 9:6). This verse, along with others in the Bible, provides a clear depiction of the hydrologic cycle long before it was fully understood by science in the 17th century.

The Hydrologic Cycle in the Bible

Amos 9:6: "He…calls for the waters of the sea, and pours them out on the face of the earth…" This verse succinctly captures the essence of the hydrologic cycle, describing the continuous movement of water from the sea to the earth and back again. The ancient understanding embedded in this verse highlights a cycle of evaporation, precipitation, and return flow that modern science has only detailed in recent centuries.

Ecclesiastes 1:7: "All the rivers run into the sea, yet the sea is not full; to the place from which the rivers come, there they return again." This observation by the writer of Ecclesiastes encapsulates the process of rivers feeding into the sea, which then undergoes evaporation, forming clouds that eventually precipitate and refill the rivers—a continuous and balanced cycle.

Psalm 135:7: "He causes the vapors to ascend from the ends of the earth; He makes lightning for the rain." Here, the psalmist describes the evaporation process ("vapors to ascend") and the atmospheric dynamics leading to rain, indicating an understanding of the transition from vapor to precipitation.

Ecclesiastes 11:3: "If the clouds are full of rain, they empty themselves upon the earth." This verse highlights the culmination of the cycle, where clouds, heavy with water vapor, release rain onto the earth, completing the cycle of water movement.

The Science Behind the Cycle

The hydrologic cycle is a continuous process by which water moves through the environment. It involves several key steps:

1. Evaporation: Water from oceans, rivers, and lakes evaporates due to the heat from the sun, turning into water vapor.

2. Condensation: The water vapor rises and cools in the atmosphere, forming clouds through condensation.

3. Precipitation: When clouds become saturated, they release water back to the earth as rain, snow, sleet, or hail.

4. Collection: The precipitated water collects in bodies of water such as rivers, lakes, and oceans, and the cycle begins anew.

The Role of the Mississippi River: The Mississippi River alone discharges over six million gallons of water per second into the Gulf of Mexico. This vast amount of water, along with contributions from countless other rivers, enters the cycle through evaporation. The water vapor forms clouds that travel over the land, eventually precipitating and continuing the hydrologic cycle.

The Hydrologic Cycle and Divine Design

The intricate and balanced nature of the hydrologic cycle reflects a system of divine design, as outlined in the Bible. The continuous movement of water not only sustains life but also maintains ecological balance. The biblical descriptions underscore the wisdom and foresight embedded in the Scriptures, revealing insights into natural processes that are fundamental to life on earth.

The Mississippi River Example: The vast quantities of water flowing into the sea from the Mississippi River and other rivers are a testament to this divinely orchestrated system. Despite the enormous influx, the seas do not overflow, thanks to the balanced process of evaporation and precipitation described in the Scriptures.

Biblical Insights and Modern Science

The Scriptures offer a profound alignment with modern scientific understanding of the hydrologic cycle. This harmony between ancient texts and contemporary science underscores the timeless relevance and accuracy of biblical wisdom. The descriptions in Amos, Ecclesiastes, and Psalms provide a clear depiction of water's journey, reflecting an advanced understanding of natural phenomena.

Practical Implications: Understanding the hydrologic cycle is crucial for managing water resources, predicting weather patterns, and addressing environmental challenges.

The Bible's early references to these processes highlight the importance of water management and the need for stewardship of this vital resource.

The hydrologic cycle, as described in the Bible, reveals a deep understanding of natural processes that are essential for sustaining life on earth. The verses from Amos, Ecclesiastes, and Psalms illustrate the continuous movement of water and its critical role in the environment. This ancient wisdom not only aligns with modern scientific discoveries but also emphasizes the divine design underlying the natural world. Recognizing this connection enhances our appreciation of both the Scriptures and the intricate systems that sustain life, inspiring a sense of awe and responsibility towards our natural resources.

By integrating biblical references with scientific understanding, we gain a comprehensive view of the hydrologic cycle, appreciating its complexity and the foresight of ancient biblical writers. This holistic perspective encourages both spiritual and practical engagement with the natural world, fostering a deeper connection to the divine wisdom that guides our understanding of creation.

Amos 9:6 - Expository Study and Comprehensive Commentary

Verse (KJV): "It is he that buildeth his stories in the heaven, and hath founded his troop in the earth; he that calleth for the waters of the sea, and poureth them out upon the face of the earth: The Lord is his name."

Context and Summary

Amos 9:6 is part of the closing chapter of the Book of Amos, where the prophet Amos delivers a final message of judgment and hope from God to the Israelites. In this verse, Amos emphasizes God's supreme power and authority over the natural world and His role as the Creator and Sustainer of the universe.

Interpretation and Commentary

"It is he that buildeth his stories in the heaven": This phrase refers to God's architectural power, describing Him as the builder of the heavens. The term "stories" (or "upper chambers") signifies the different layers or levels of the heavens, indicating the grandeur and complexity of God's creation.

References:

- Psalm 104:3: "Who layeth the beams of his chambers in the waters: who maketh the clouds his chariot: who walketh upon the wings of the wind."

- Job 22:12-14: "Is not God in the height of heaven? and behold the height of the stars, how high they are!... and he walketh in the circuit of heaven."

"And hath founded his troop in the earth": This highlights God's establishment of the earth. The "troop" can be interpreted as God's army or the multitude of His creations, indicating the ordered and purposeful nature of His work on earth.

References:

- Isaiah 40:22: "It is he that sitteth upon the circle of the earth, and the inhabitants thereof are as grasshoppers; that stretcheth out the heavens as a curtain, and spreadeth them out as a tent to dwell in."

- Psalm 104:5: "Who laid the foundations of the earth, that it should not be removed for ever."

"He that calleth for the waters of the sea, and poureth them out upon the face of the earth": This phrase depicts the hydrologic cycle, showing God's control over the natural processes. He commands the waters to rise (evaporation) and then to descend (precipitation), sustaining life on earth.

References:

- Jeremiah 10:13: "When he uttereth his voice, there is a multitude of waters in the heavens, and he causeth the vapours to ascend from the ends of the earth; he maketh lightnings with rain, and bringeth forth the wind out of his treasures."

- Ecclesiastes 1:7: "All the rivers run into the sea; yet the sea is not full; unto the place from whence the rivers come, thither they return again."

"The Lord is his name": This closing declaration emphasizes that all these mighty acts are done by Yahweh, the covenant God of Israel, underscoring His sovereignty and power.

References:

- Exodus 15:3: "The Lord is a man of war: the Lord is his name."

- Jeremiah 33:2: "Thus saith the Lord the maker thereof, the Lord that formed it, to establish it; the Lord is his name."

Concordance and Key Themes

- Sovereignty of God: God's control over the heavens and the earth, as well as natural processes, highlights His supreme authority and sovereignty.

- Creation and Sustenance: The verse emphasizes God's role as the Creator and Sustainer, responsible for the continuous functioning of the natural world.

- Hydrologic Cycle: The mention of calling for the waters of the sea and pouring them upon the earth reflects an understanding of the water cycle, showing the Bible's insight into natural processes.

- Divine Name: The proclamation "The Lord is his name" serves as a powerful reminder of God's identity and covenant relationship with His people.

Theological Insights

Amos 9:6 provides a profound theological perspective on God's omnipotence and omnipresence. It reinforces the belief that God is intricately involved in the workings of the universe, governing both the heavens and the earth. This verse calls believers to recognize and revere God's mighty acts and to trust in His continued provision and governance.

Practical Application

Understanding the sovereignty of God as depicted in Amos 9:6 can inspire believers to:

- Trust in God's Control: In times of uncertainty, remembering God's absolute control over creation can provide comfort and assurance.

- Revere God's Majesty: Reflecting on the grandeur of God's creation can deepen one's worship and awe of Him.

- Stewardship of Creation: Recognizing God's meticulous design in nature can motivate responsible stewardship and care for the environment.

By examining Amos 9:6 through various lenses—context, interpretation, references, concordance, and application—we gain a deeper appreciation of its rich theological and practical significance.

CHAPTER 16

THE SCIENCE OF OCEANOGRAPHY

Psalm 8:8: "…and the fish of the sea that pass through the paths of the seas."

Oceanography, the study of the physical and biological aspects of the ocean, has unveiled numerous mysteries of the deep sea, revealing the complex systems and pathways that traverse our vast oceans. One of the most intriguing aspects of this field is the discovery of ocean currents, which act as the "paths of the seas." Remarkably, the Bible referenced these paths thousands of years before their scientific discovery, particularly in Psalm 8:8. This chapter explores the intersection of ancient scripture and modern science, focusing on the life and work of Matthew Maury, who is often regarded as the father of oceanography.

The Biblical Insight

Psalm 8:8 states, "…and the fish of the sea that pass through the paths of the seas." At first glance, this verse seems purely poetic, describing the wonders of God's creation. However, upon closer inspection, it contains profound scientific truth. The "paths of the seas" refer to the ocean currents, which are vast, powerful flows of water that move through the world's oceans.

Historical Context

The understanding of the ocean as a dynamic system with currents and pathways was not established until the mid-19th century. Before this period, the ocean was largely seen as a vast, uncharted expanse. The discovery of these currents revolutionized maritime navigation, weather forecasting, and our overall understanding of marine ecosystems.

Matthew Maury: The Pioneer of Oceanography

Matthew Fontaine Maury (1806–1873), a United States Navy officer, was fascinated by the Bible's reference to the "paths of the seas." His belief in the accuracy of scripture led him to investigate these oceanic pathways scientifically. Maury meticulously gathered data from ship logs, maps, and weather reports, identifying patterns that would later be recognized as ocean currents.

Maury's Contributions

1. Mapping Ocean Currents: Maury's work culminated in the publication of "The Physical Geography of the Sea" in 1855, which is often regarded as the first textbook on oceanography. His maps detailed the major ocean currents, including the Gulf Stream, which he famously charted, enhancing the efficiency and safety of sea travel.

2. Maritime Navigation: By understanding and utilizing these currents, sailors could significantly reduce travel time and fuel consumption. Maury's charts became indispensable tools for navigators, transforming maritime trade and exploration.

3. Meteorology and Climate: Maury's studies also linked ocean currents to weather patterns, laying the groundwork for modern meteorology. He recognized the impact of these currents on global climate, a concept that remains crucial in contemporary climate science.

Ocean Currents: The Modern Understanding

Ocean currents are driven by several factors, including wind patterns, the Earth's rotation (Coriolis effect), temperature differences, and salinity gradients. These currents play a critical role in regulating the Earth's climate, distributing heat from the equator to the poles and influencing weather systems.

Major Ocean Currents

1. The Gulf Stream: This powerful, warm Atlantic Ocean current originates in the Gulf of Mexico and flows up the eastern coast of the United States before crossing the Atlantic towards Europe. It has a significant impact on the climate of the eastern seaboard and Western Europe.

2. The California Current: A cold Pacific Ocean current that moves southward along the western coast of North America, influencing the climate and marine life of the region.

3. The Antarctic Circumpolar Current: The largest ocean current, which flows clockwise around Antarctica, plays a crucial role in global ocean circulation.

The Intersection of Faith and Science

The discovery of ocean currents as described in the Bible underscores the harmony between faith and science. Maury's work exemplifies how scientific inquiry can be inspired by scriptural truths, leading to profound discoveries that benefit humanity. His faith-driven exploration and subsequent scientific achievements highlight the potential for collaboration between religious belief and scientific endeavor.

The science of oceanography, with its roots in ancient scripture and its development through the pioneering work of Matthew Maury, demonstrates the enduring value of integrating faith with scientific inquiry. Psalm 8:8's reference to the "paths of the seas" prefigured the discovery of ocean currents by thousands of years, showcasing the Bible's profound insights into the natural world. Today, oceanography continues to unravel the mysteries of the seas, driven by a legacy of curiosity, faith, and scientific rigor.

By understanding these "paths of the seas," we gain not only navigational and climatic insights but also a deeper appreciation for the interconnectedness of our planet's systems, a testament to the wisdom embedded in the ancient texts and the ongoing quest for knowledge.

Psalm 8:8

"The fowl of the air, and the fish of the sea, and whatsoever passeth through the paths of the seas." (KJV)

Verse Analysis

This verse is part of Psalm 8, a hymn of praise to God for His majestic creation and His special care for humanity. The psalmist marvels at the grandeur of the universe and humanity's privileged place within it. Verse 8 specifically mentions God's creation of birds, fish, and other creatures that inhabit the seas and move through its pathways.

King James Bible's References

1. "The fowl of the air": This phrase refers to birds and is a common biblical expression used to describe

creatures that fly or dwell in the sky (Genesis 1:20; Deuteronomy 4:17).

2. "The fish of the sea": This phrase refers to the creatures that inhabit the waters, specifically fish. It emphasizes the diversity and abundance of marine life (Genesis 1:20-21; Jonah 1:17).

3. "Whatsoever passeth through the paths of the seas": This phrase suggests the movement of various sea creatures through the oceanic currents or pathways. It speaks to the vastness and interconnectedness of the seas (Job 28:8; Isaiah 43:16).

Interpretation and Commentary

1. God's Sovereignty over Creation: This verse reflects the psalmist's recognition of God's sovereignty and creative power over all living creatures. The mention of birds, fish, and sea creatures highlights God's diverse and intricate creation.

2. God's Provision for His Creatures: The phrase "paths of the seas" indicates that God has established specific pathways or currents in the oceans through which sea creatures travel. This demonstrates God's provision and care for His creation, ensuring that even the smallest sea creatures have a means of navigation and sustenance.

3. Humanity's Role as Stewards: As part of God's creation, humanity is called to be stewards of the earth and its

creatures (Genesis 1:26). This verse reminds us of our responsibility to care for God's creation and to appreciate the beauty and complexity of the natural world.

4. Scientific Insight: While the psalmist likely did not have a scientific understanding of ocean currents, this verse has been noted for its prescience regarding the paths of the seas. Modern oceanography has confirmed the existence of ocean currents that serve as highways for marine life, echoing the psalmist's recognition of these "paths."

Concordance

- "Paths": The Hebrew word used here is "mesillah," which can also mean a highway or a path. It suggests a defined route or course, indicating the order and design in the movement of sea creatures (Psalm 107:7; Proverbs 8:20).

- "Seas": The Hebrew word "yam" is used to refer to the seas or large bodies of water. It signifies the vastness and expansiveness of the oceans, which God has created and sustains (Genesis 1:10; Exodus 14:21).

Psalm 8:8 is a testament to the psalmist's awe and wonder at God's creation, specifically the intricate design and order found in the animal kingdom. It invites us to contemplate the beauty and complexity of the natural world and to recognize God's handiwork in all living creatures, from the birds of the air to the fish of the sea.

CHAPTER 17

THE ORIGIN OF SEXES

In the book of Matthew, Jesus stated, "He who made them at the beginning 'made them male and female.'" This fundamental truth about the nature of sexes has profound implications for our understanding of life's origins.

Almost all forms of complex life on Earth exhibit sexual dimorphism, meaning they have distinct male and female sexes. This is true for mammals like horses and dogs, as well as fish, insects, primates, and many other organisms. The existence of two sexes is essential for reproduction in these species, as each sex plays a unique role in the process.

But if we consider the theory of evolution, which posits that life forms gradually evolved over time, we encounter a puzzling question: which sex came first? According to evolutionary theory, the emergence of distinct sexes would have been a gradual process. However, this raises further questions about how each sex could have functioned and reproduced independently before the other sex evolved.

If, for instance, males evolved before females, how did these early males reproduce without females? Conversely, if females evolved first, how did they reproduce without males? The idea that each sex evolved independently and then somehow converged to form complementary reproductive systems is difficult to reconcile with the principles of evolutionary biology.

Moreover, if each sex was capable of reproducing independently, why would they have developed complex, interdependent reproductive systems that require the presence of the other sex for successful reproduction? Evolutionary theory struggles to provide a satisfactory explanation for the simultaneous and interdependent emergence of male and female sexes.

In contrast, the biblical account offers a coherent explanation for the origin of the sexes. It suggests that from the beginning, God created male and female as complementary counterparts, each uniquely designed for the other. This harmonious design reflects God's intention for reproduction and the continuation of life on Earth.

The existence of two sexes, each with its own reproductive system, speaks to the wisdom and purpose of a Creator who designed life with intention and purpose. Far from being a product of random chance or evolutionary processes, the origin of sexes points to a deliberate and intelligent design that reflects the nature of God Himself.

As we ponder the intricacies of sexual reproduction and the complementary nature of male and female, let us marvel at the wisdom and foresight of the Creator who made them male and female, from the beginning.

Matthew 19:4 (KJV): "And he answered and said unto them, Have ye not read, that he which made them at the beginning made them male and female,"

In this verse, Jesus responds to the Pharisees' question about divorce by referring to the creation account found in Genesis. Let's break down this verse with a comprehensive study and commentary.

King James Bible's References:

- This verse alludes to Genesis 1:27 and Genesis 5:2, where it is stated that God created mankind in His own image, male and female He created them.

Interpretation:

1. Jesus' Authority: Jesus begins His response to the Pharisees' question by reminding them of the authority of

Scripture. He asks them if they have not read the creation account in Genesis, implying that the answer to their question can be found in the foundational teachings of Scripture.

2. Creation of Male and Female: Jesus refers to the creation of male and female as a foundational principle established by God "at the beginning." This indicates that the differentiation between male and female was not a later development but was present from the very outset of creation.

Commentary:

- God's Design for Marriage: By referencing the creation of male and female, Jesus reaffirms God's original design and intention for marriage. In Genesis, God created Eve as a suitable companion for Adam, establishing the foundation for marriage as a union between one man and one woman.

- Equality and Complementarity: The fact that God created both male and female reflects the equality and complementarity of the sexes. Both men and women are created in the image of God and have inherent value and dignity. At the same time, their differences highlight the complementary nature of the sexes, each bringing unique qualities and perspectives to relationships and society.

- Response to Divorce Question: Jesus' reference to the creation of male and female serves as the basis for His subsequent teaching on divorce. By appealing to God's original design for marriage, Jesus underscores the permanence and sanctity of the marital bond, emphasizing that divorce was not part of God's original plan.

Concordance:

- The concept of male and female being created by God "at the beginning" is foundational to biblical anthropology and theology, providing the basis for understanding human nature, relationships, and the institution of marriage.

In summary, Matthew 19:4 encapsulates Jesus' affirmation of God's original design for marriage and human relationships, rooted in the creation of male and female. This verse highlights the importance of recognizing and honoring God's intentions as revealed in Scripture, particularly in matters concerning marriage and family life.

CHAPTER 18

COUNTLESS STARS

In the vast expanse of the heavens, beyond the reaches of our imagination, lie the countless stars. Jeremiah, an ancient prophet, spoke of this vastness over 2,500 years ago, proclaiming, "As the host of heaven cannot be numbered, nor the sand of the sea measured…" (Jeremiah 33:22).

At the time these words were penned, the understanding of the universe was limited. The number of observable stars was fewer than 1,100, as recorded by Ptolemy in The Almagest. Yet, even then, Jeremiah spoke of a reality that transcended human comprehension. He spoke of a multitude so vast that it could not be quantified, likening it to the endless grains of sand on the seashore.

Today, with the advancement of science and technology, we have gained a greater understanding of the universe. We now know that there are billions upon billions of stars in the observable universe alone, estimated to be around 10^25 stars. And yet, even with this vast number, we still cannot truly comprehend the full extent of the universe. The universe continues to expand, revealing even more wonders that lie beyond our current understanding.

The vastness of the stars reminds us of the greatness of the Creator. It speaks to the awe-inspiring nature of God, who spoke these stars into existence. It reminds us of our place in the universe, as tiny specks in a vast cosmic sea. And yet, it also reminds us of the importance and significance that each of us holds in the eyes of our Creator.

As we gaze upon the countless stars in the sky, let us be reminded of the greatness of God and the wonders of His creation. Let us be humbled by the vastness of the universe and the mysteries that lie beyond our grasp. And let us be inspired to seek out the truths that lie hidden in the depths of the cosmos, knowing that the God who created the stars also holds the answers to life's deepest questions.

Jeremiah 33:22 (KJV): "As the host of heaven cannot be numbered, neither the sand of the sea measured: so will I multiply the seed of David my servant and the Levites that minister unto me."

This verse is a proclamation from God, delivered through the prophet Jeremiah. It begins with a comparison between the countless stars in the sky and the immeasurable grains of sand on the seashore. Just as these elements of nature are beyond human ability to count or measure, so too will God multiply the descendants of David and the Levites who serve Him.

Interpretation:

1. Host of Heaven: This phrase refers to the stars in the sky, symbolizing the vastness and uncountable nature of God's creation.

2. Seed of David: This refers to the descendants of King David, indicating that God will multiply his lineage. This promise is often understood in a spiritual sense, pointing to the Messiah who would come from the line of David, namely Jesus Christ.

3. Levites: The Levites were a tribe of Israel set apart for the service of the Tabernacle and later the temple. God promises to increase their number as well, indicating a continuation and expansion of their ministry.

Commentary:

This verse is part of a larger passage where God promises to restore and bless His people after a period of exile and judgment. The comparison to the stars and the sand emphasizes the abundance and greatness of God's blessings. Despite the current challenges and hardships faced by the Israelites, God assures them of a future filled with abundance and prosperity.

Concordance:

- Stars: Genesis 15:5, Judges 5:20, Isaiah 40:26

- Sand of the Sea: Genesis 22:17, Joshua 11:4, Jeremiah 15:8

- Multiply: Genesis 22:17, Exodus 32:13, Deuteronomy 1:10

CHAPTER 19

THE VITAL IMPORTANCE OF BLOOD

Leviticus 17:11 (Written 3,500 Years Ago): "For the life of the flesh is in the blood."

In the annals of ancient wisdom, amidst the sacred texts of the Judeo-Christian tradition, there exists a profound declaration: "For the life of the flesh is in the blood." These words, spoken millennia ago, encapsulate a truth that modern science has only begun to unravel—the indispensable role of blood in sustaining life itself.

Throughout history, humanity has grappled with the mystery of blood, often attributing to it an almost mystical significance. From the ancient practice of bloodletting to the intricate study of hematology in contemporary medicine, the understanding of blood has evolved, revealing its profound importance in the intricate dance of life.

In the bygone era of medical ignorance, the sick were often subjected to the barbaric practice of bloodletting, believed to purge the body of illness. However, as the pages of time turned, the true nature of blood began to emerge from the shadows of superstition.

Today, we stand at the threshold of enlightenment, armed with the knowledge that blood is not merely a crimson fluid coursing through our veins, but the very essence of vitality itself. It serves as the vehicle for the transportation of vital nutrients and oxygen, ferrying them to every cell in our bodies. Like a diligent courier, it removes waste materials, ensuring the cleanliness and efficiency of our cellular machinery.

Moreover, blood plays a pivotal role in regulating body temperature, maintaining the delicate balance that sustains life. It is the silent conductor orchestrating the symphony of bodily functions, ensuring harmony amidst the chaos of existence.

Yet, despite its resilience and adaptability, blood is a finite resource—a truth starkly illustrated by the ominous specter of blood loss. In the face of hemorrhage, the life-giving essence drains away, leaving behind only the cold embrace of death.

In light of these revelations, we are compelled to acknowledge the sanctity of blood, to treat it not as a mere commodity but as the sacred wellspring from which life itself flows. We are called to honor the sacrifices made by those who have donated their blood, recognizing that each drop given is a gift of immeasurable value—a gift that sustains, heals, and restores.

Let us, therefore, heed the ancient wisdom inscribed in the sacred texts, and cherish the blood that courses through our veins. Let us safeguard its sanctity, for in doing so, we honor the very essence of life itself.

The Life-Giving Properties of Blood

Blood plays a crucial role in maintaining the body's functions. It serves as a transportation system, delivering oxygen and nutrients to cells while removing waste products. The circulatory system, powered by the heart, ensures that every part of the body receives the essential elements it needs to survive and thrive.

Historical Perspectives on Blood

Throughout history, blood has been a symbol of life and vitality, as well as a source of mystery and fear. Ancient civilizations, including the Israelites, Egyptians, Greeks, and

Romans, recognized the importance of blood in rituals and sacrifices. Bloodletting, the practice of deliberately bleeding a person, was a common medical treatment believed to balance the body's humor.

Scientific Understanding of Blood

Modern science has deepened our understanding of blood, revealing its complex composition and functions. Blood is primarily made up of plasma, red blood cells, white blood cells, and platelets. Each component plays a vital role in maintaining health, from fighting infections to clotting wounds.

Spiritual Significance of Blood

In many religious and spiritual traditions, blood holds symbolic significance. In Christianity, for example, the shedding of Jesus Christ's blood is seen as an atoning sacrifice for humanity's sins. The act of communion, where believers drink wine symbolizing Christ's blood, is a central ritual in many Christian denominations.

The words of Leviticus 17:11, written millennia ago, still resonate today, reminding us of the profound connection between blood and life. As we continue to unravel the mysteries of biology and medicine, let us not forget the awe-inspiring complexity and importance of this vital fluid that sustains us all.

CHAPTER 20

THE SIGNIFICANCE OF BLOOD CLOTTING IN CIRCUMCISION

In Genesis 17:12, God commands Abraham that every male child in his generation shall be circumcised when they are eight days old. This seemingly arbitrary timing is actually deeply significant and reflects a remarkable understanding of human physiology, specifically regarding blood clotting.

Medical science has only recently discovered that the eighth day of a newborn's life is optimal for circumcision due to the intricate process of blood clotting. This process involves various coagulating factors in the blood that work together to form a clot and stop bleeding. One crucial factor is vitamin K, which is essential for the production of several coagulation factors. Interestingly, vitamin K does not reach sufficient levels in newborns until after the seventh day of life.

On the eighth day, however, the levels of prothrombin, a key coagulation factor, reach their peak, exceeding the normal level by about 10 percent. This peak ensures that the blood has optimal clotting ability, reducing the risk of excessive bleeding during circumcision.

The fact that God specifically chose the eighth day for circumcision, long before the discovery of the intricate details of blood clotting, is a testament to His divine wisdom and knowledge of human physiology. It also highlights the importance of following God's commands, even when we may not fully understand the reasons behind them.

This example from Genesis serves as a reminder of the intricate design of the human body and the wisdom of God's commands. It also underscores the importance of trusting in God's wisdom, even when His instructions may seem puzzling to us.

As we reflect on this biblical teaching, let us marvel at the complexity of our bodies and the wisdom of our Creator, who designed us with such care and foresight. Let us also be reminded of the importance of obedience to God's commands, trusting that His ways are higher than ours, and His wisdom far surpasses our understanding.

In Genesis 17:12, God instructs Abraham regarding circumcision: "And he that is eight days old shall be circumcised among you, every man child in your generations, he that is born in the house, or bought with money of any stranger, which is not of thy seed."

King James Bible's References:

- Leviticus 12:3: "And in the eighth day the flesh of his foreskin shall be circumcised."

- Luke 2:21: "And when eight days were accomplished for the circumcising of the child, his name was called JESUS, which was so named of the angel before he was conceived in the womb."

Interpretation:

The instruction given to Abraham regarding circumcision specifies that it should be performed on males when they are eight days old. This ritual was to be carried out not only for Abraham's descendants but also for any male born in his household or acquired through other means. This practice signified the covenant between God and Abraham's descendants, symbolizing their consecration and separation unto God.

Commentary:

Circumcision was a significant covenantal act in ancient Israel, symbolizing a separation and consecration of the male child to God. The timing of circumcision on the eighth day is notable, as it coincides with the peak of vitamin K and prothrombin levels in a newborn's blood, ensuring optimal clotting ability and minimizing the risk of bleeding.

In the New Testament, we see Jesus being circumcised according to the law on the eighth day, highlighting his fulfillment of the law and his identification with God's covenant people.

Concordance:

- Circumcision (Genesis 17:12): This act symbolizes the covenant between God and Abraham's descendants. It signifies consecration and separation unto God.

- Eighth day (Genesis 17:12): The specific timing of circumcision, which coincides with the optimal levels of vitamin K and prothrombin in a newborn's blood, ensuring proper clotting ability.

This verse underscores the importance of obedience to God's commands and highlights the symbolism and significance of circumcision in the covenantal relationship between God and His people.

CHAPTER 21

THE TIMELESS WISDOM OF HYGIENE LAWS

In the book of Leviticus, amidst the laws and rituals that may seem archaic to modern readers, there lies a profound wisdom that transcends time and culture. Leviticus 15:13 provides a remarkable example of this wisdom, offering insights into hygiene practices that are remarkably relevant even in today's world.

The verse states, "And when he who has a discharge is cleansed of his discharge, then he shall count for himself seven days for his cleansing, wash his clothes, and bathe his body in running water; then he shall be clean." At first glance, this verse may appear to be simply a ritualistic prescription for purification. However, a deeper examination reveals a remarkably advanced understanding of hygiene principles that were far ahead of their time.

One of the key aspects of this verse is the emphasis on washing with running water. This simple act, which may have seemed like common sense to the ancient Israelites, held profound implications for public health. It is well-documented that until the 19th century, medical practitioners did not understand the importance of washing hands with running water. Instead, they would wash their hands in still water, unknowingly spreading germs and contributing to the spread of disease. It was only through the pioneering work of individuals like Dr. Ignaz Semmelweis that the importance of hand hygiene in preventing the spread of disease became widely recognized.

Dr. Semmelweis, working in 19th-century Vienna, observed a high mortality rate among women giving birth in hospitals. He discovered that this was largely due to doctors not washing their hands between examinations, leading to the transmission of infectious diseases. When he implemented a policy of handwashing with chlorinated lime water, the mortality rate dropped dramatically, highlighting the life-saving impact of this seemingly simple practice.

The wisdom of Leviticus 15:13 extends beyond the specific act of handwashing to encompass broader principles of hygiene. The verse emphasizes the importance of cleanliness in preventing the spread of disease, a concept that remains central to modern public health practices. It also underscores the idea of personal responsibility for hygiene, with individuals being instructed to cleanse themselves and their belongings.

In conclusion, while the hygiene laws of Leviticus may have been given thousands of years ago, their relevance and wisdom endure to this day. They remind us of the importance of simple yet effective practices like handwashing in maintaining health and preventing disease. They also serve as a testament to the timeless wisdom of the Bible, which continues to offer insights and guidance for living a healthy and fulfilling life.

Leviticus 15:13 (KJV): "And when he that hath an issue is cleansed of his issue; then he shall number to himself seven days for his cleansing, and wash his clothes, and bathe his flesh in running water, and shall be clean."

This verse is part of a larger passage in Leviticus 15 that deals with various bodily discharges and the laws surrounding them. The context is important for understanding the significance of these laws in the cultural and religious context of ancient Israel. Here, the focus is on the cleansing rituals required after a person with a discharge has been healed.

Interpretation:

1. Cleansing Period: The individual is instructed to count seven days from the time they are healed. This period likely served both a practical and symbolic purpose, allowing time for complete healing and also symbolizing a complete cycle of purification.

2. Washing Clothes: Washing clothes was a common practice for purification in ancient cultures, symbolizing a fresh start and the removal of impurity.

3. Bathing in Running Water: The use of running water for bathing was significant, as stagnant water was often associated with impurity. Running water was considered purifying and was likely more effective in removing physical impurities.

4. Symbolism of Cleanliness: Beyond the physical act of washing, these rituals symbolized the importance of spiritual and moral cleanliness. They reinforced the idea that one must be clean before approaching the sacred or rejoining the community.

5. Emphasis on Personal Responsibility: The individual is instructed to perform these rituals themselves,

emphasizing personal responsibility for maintaining cleanliness and purity.

Commentary:

The laws regarding bodily discharges in Leviticus 15 are often viewed as archaic or irrelevant in modern times. However, they served a crucial role in ancient Israelite society, emphasizing the importance of cleanliness and purity before God and within the community. While the specific rituals may no longer apply, the underlying principles of personal hygiene, cleanliness, and respect for the sacred remain relevant today.

Concordance:

This verse emphasizes the importance of cleanliness and purification rituals in ancient Israelite culture. It highlights the symbolic and practical aspects of these rituals and their significance in maintaining spiritual and physical purity.

CHAPTER 22

LAWS OF QUARANTINE

In ancient times, long before the advent of modern medicine, the Bible provided instructions on how to deal with infectious diseases. Leviticus 13:46, written over 3,500 years ago, outlines a key principle: "All the days he has the sore he shall be unclean. He is unclean, and he shall dwell alone; his dwelling shall be outside the camp."

The significance of this verse becomes apparent when considering the historical context. In 1490 BC, when these laws were given, the concept of quarantine was not a common practice. It wasn't until the seventeenth century that modern man began to understand and implement the idea of isolating individuals with contagious diseases.

During the Black Death of the fourteenth century, for example, those who were sick or dead were often kept in the same rooms as the rest of the family. The lack of understanding about infectious diseases led people to attribute these epidemics to "bad air" or "evil spirits."

However, had they followed the biblical instructions found in Leviticus, many lives could have been saved. The meticulous laws outlined in Leviticus 13 regarding leprosy can be seen as the first model of sanitary legislation. Arturo Castiglione, in his work A History of Medicine, acknowledges the importance of these biblical laws, stating that they represent the foundation of sanitary legislation.

The concept of quarantine as a means to prevent the spread of disease is not just a historical curiosity; it remains relevant today. In the face of global pandemics, such as the COVID-19 outbreak, the principles of isolation and quarantine continue to play a crucial role in controlling the spread of infectious diseases.

The Bible's laws of quarantine serve as a reminder of the wisdom and foresight contained in its pages. They demonstrate that even in matters of public health, the ancient wisdom of Scripture can provide valuable insights that are still applicable in the modern world.

Certainly! Here is an expository study and comprehensive commentary on Leviticus 13:46, using the King James Bible's references, interpretation, commentary, and concordance:

Leviticus 13:46 (KJV): "All the days wherein the plague shall be in him he shall be defiled; he is unclean: he shall dwell alone; without the camp shall his habitation be."

Verse Analysis:

This verse concludes the chapter on leprosy in Leviticus, summarizing the isolation requirements for those afflicted with the disease. It emphasizes the need for separation from the community to prevent the spread of contagion.

Interpretation:

1. Defilement: The person with leprosy was considered ritually unclean, emphasizing the spiritual and social implications of the disease.

2. Isolation: The individual was required to dwell alone, highlighting the physical separation from the community to prevent contamination.

3. Location: Their habitation was to be outside the camp, indicating a removal from the sacred space of the community.

Commentary:

- The laws regarding leprosy in Leviticus 13 demonstrate God's concern for the physical, spiritual, and social well-being of His people. They were designed not only to prevent the spread of disease but also to teach lessons about holiness, separation, and community welfare.

- The requirement for isolation may seem harsh, but it served a crucial purpose in protecting the community from the spread of contagious diseases, demonstrating the importance of public health in biblical times.

- This verse also reflects the theological concept of purity and impurity in the Old Testament, where physical conditions like leprosy were seen as outward signs of spiritual impurity and required specific rituals for cleansing and restoration.

Concordance:

- Defilement: Leviticus 13:3, 44-45; Numbers 5:2; Ezekiel 44:25.

- Unclean: Leviticus 13:3, 11, 14, 15, 20, 22, 25, 27, 30, 36, 44; Numbers 5:2; 2 Chronicles 26:21; Isaiah 52:11; Ezekiel 44:23; Matthew 8:2.

- Dwell alone: Leviticus 13:46; Numbers 5:2.

- Without the camp: Leviticus 13:46; Numbers 5:3; 12:14-15; Hebrews 13:12.

In conclusion, Leviticus 13:46 provides valuable insights into ancient practices regarding contagious diseases and emphasizes the importance of public health and spiritual purity in the community.

CHAPTER 23

THE DISAPPEARING DINOSAUR

The mystery of the disappearance of dinosaurs has captivated scientists and researchers for centuries. These colossal creatures, once rulers of the Earth, vanished without a trace, leaving behind only their fossilized remains. Modern science has proposed various theories to explain their extinction, from asteroid impacts to climate change. However, an ancient text, the Bible, may provide a unique perspective on this enigma.

In the Book of Job, written over 3,500 years ago, there is a passage that describes a creature called the behemoth. This beast is portrayed as the largest of all creatures, with immense strength and habitat among the trees. It is herbivorous, with bones as strong as bronze and iron. It stands unmoved in the midst of a raging river and is impervious to snares. The passage concludes by stating, "Only He who made him can bring near His sword."

Could the behemoth be a description of the dinosaurs? And does this passage offer a clue to their disappearance?

The Behemoth: A Dinosaur?

The description of the behemoth in Job bears striking similarities to the characteristics of dinosaurs. It is described as the largest of all creatures, fitting the description of many dinosaur species. Its herbivorous diet aligns with the fossil evidence of many dinosaurs being plant-eaters. The reference to its tail being like a cedar and its bones like bronze and iron could be metaphorical, describing its immense size and strength.

Furthermore, the behemoth's habitat among the trees and its ability to stand unmoved in a raging river suggest a creature of colossal proportions, reminiscent of the giant sauropods that once roamed the Earth. The reference to being impervious to snares could indicate its invulnerability to human threats, as only its Creator could bring about its end.

The Disappearance of the Dinosaurs

The most intriguing part of the passage is the statement, "Only He who made him can bring near His sword." This implies that the behemoth, and by extension, the dinosaurs, could only be threatened by their Creator. This raises the question: Did God cause the extinction of the dinosaurs?

While the Bible does not explicitly state that God caused the extinction of the dinosaurs, it does suggest that their disappearance was within His power. If the behemoth indeed represents the dinosaurs, then their extinction may have been part of God's plan.

The mystery of the disappearing dinosaur continues to puzzle scientists and researchers. While modern science offers various theories, the ancient text of the Bible provides a unique perspective. The description of the behemoth in the Book of Job aligns closely with the characteristics of dinosaurs, suggesting that their extinction may have been orchestrated by their Creator. Whether this is the definitive answer to the mystery remains a subject of debate. Nonetheless, the Bible's account offers a thought-provoking insight into the fate of these ancient giants.

Job Chapter 40:15-24 - The Behemoth

Verse 15-16: "Behold now behemoth, which I made with thee; he eateth grass as an ox. Lo now, his strength is in his loins, and his force is in the navel of his belly."

King James Bible (KJV) References:

- Behemoth: This is a Hebrew word that means "beast" or "large animal." It is often interpreted to refer to a hippopotamus or an elephant.

- Grass: Indicates that the behemoth is a herbivore.

- Strength in his loins: Indicates the creature's physical power.

Interpretation and Commentary:

- The behemoth is described as a large, herbivorous animal, possibly a hippopotamus or an elephant.

- Its strength is in its loins and belly, indicating its physical power and vitality.

Verse 17-18: "He moveth his tail like a cedar: the sinews of his stones are wrapped together. His bones are as strong pieces of brass; his bones are like bars of iron."

King James Bible (KJV) References:

- Tail like a cedar: Indicates a large, strong tail.

- Sinews of his stones: Refers to the muscles around its testicles, symbolizing its reproductive strength.

- Bones like brass and iron: Emphasizes the creature's strength and durability.

Interpretation and Commentary:

- The behemoth's tail is compared to a cedar tree, highlighting its size and strength.

- The description of its reproductive organs indicates its vitality and ability to reproduce.

- Its bones are likened to brass and iron, emphasizing their strength and durability.

Verse 19-20: "He is the chief of the ways of God: he that made him can make his sword to approach unto him. Surely the mountains bring him forth food, where all the beasts of the field play."

King James Bible (KJV) References:

- Chief of the ways of God: Suggests that the behemoth is among the most impressive of God's creations.

- Mountains bring him forth food: Indicates the behemoth's habitat and diet.

Interpretation and Commentary:

- The behemoth is considered one of the greatest of God's creations, highlighting its magnificence.

- It is portrayed as a creature of the mountains, where it finds its food and lives among other wild animals.

Verse 21-23: "He lieth under the shady trees, in the covert of the reed, and fens. The shady trees cover him with their shadow; the willows of the brook compass him about.

Behold, he drinketh up a river, and hasteth not: he trusteth that he can draw up Jordan into his mouth."

King James Bible (KJV) References:

- Lies under shady trees: Describes the creature's habitat, indicating it seeks shade.

- Drinketh up a river: Suggests the behemoth can consume large amounts of water.

Interpretation and Commentary:

- The behemoth is depicted as living among shady trees and reeds, indicating a lush, riverside habitat.

- It is said to be able to drink large quantities of water without fear, showing its confidence and dominance in its environment.

Verse 24: "He taketh it with his eyes: his nose pierceth through snares."

King James Bible (KJV) References:

- Taketh it with his eyes: Suggests the behemoth is unafraid and confident.

- Nose pierceth through snares: Indicates the creature is not easily caught or trapped.

Interpretation and Commentary:

- The behemoth is described as fearless and untroubled by dangers, including traps.

Summary:

The passage describes the behemoth as a powerful and impressive creature, possibly a hippopotamus or an elephant, with great physical strength and vitality. It inhabits a lush, riverside environment and is unafraid of dangers, including traps. The passage emphasizes the behemoth's magnificence and suggests it is among the most impressive of God's creations.

SHIMBA
PUBLISHING

www.ingramcontent.com/pod-product-compliance
Lightning Source LLC
Chambersburg PA
CBHW060923140726
47996CB00001B/363